# GREEN CHEMICAL PROCESSING AND SYNTHESIS

Edited by **Iyad Karamé** and **Hassan Srour**

**Green Chemical Processing and Synthesis**
http://dx.doi.org/10.5772/65562
Edited by Iyad Karamé and Hassan Srour

**Contributors**

Noel Peter Bengzon Tan, Cheng Hao Lee, Kohki Ebitani, Shun Nishimura, Ratnaningsih Sardjono, Rahmi Rachmawati, Nikolay Britun, Tiago Silva, Guoxing Chen, Thomas Godfroid, Rony Snyders, Marie-Paule Delplancke-Ogletree, Min Lin, Changjiu Xia, Bin Zhu, Xingtian Shu, YiBin Luo, Xinxin Peng, Baorong Wang, Yao Zhang

**Published by InTech**

Janeza Trdine 9, 51000 Rijeka, Croatia



**Publishing Process Manager** Nina Kalinic
**Technical Editor** SPi Global
**Cover** InTech Design team

First published July, 2017
Printed in Croatia
Legal deposit, Croatia: National and University Library in Zagreb

Additional hard copies can be obtained from orders@intechopen.com

Green Chemical Processing and Synthesis, Edited by Iyad Karamé and Hassan Srour
p. cm.
Print ISBN 978-953-51-3259-2
Online ISBN 978-953-51-3260-8

For EU product safety concerns:
IN TECH d.o.o., Prolaz Marije Krucifikse Kozulić 3, 51000 Rijeka, Croatia,
info@intechopen.com or visit our website at intechopen.com.

# Meet the editors

**Iyad Karamé, PhD** is a Full-time Professor at the Faculty of Sciences in the Lebanese University in Beirut, and the Director of the "Laboratory of Catalysis Organometallic and Materials (LCOM)," Department of Chemistry. He got his PhD degree from the Claude Bernard Lyon 1 University in France in January 2004. He was an Assistant Professor and Researcher (ATER) at the Ecole Normale Supérieure de Lyon, France, for 1 year (2004–2005), an invited researcher at the Leibniz-Institut für Katalyse in Rostock, Germany (2005–2006), and then at the Laboratory of Organometallic Chemistry of Surface, CPE Lyon, France, till 2008. His principal axes of research are organometallic and green catalysis and organic synthesis for different applications (chelating macrocycles, ligands for metal complexes, CO2 and glycerol valorization, organocatalysts).

Hassan Srour, PhD is a research and development engineer at the École supérieure de chimie physique électronique de Lyon (C2P2-CPE de Lyon). He got his PhD degree from University of Claude Bernard Lyon 1 in France in October 2013. He was a postdoctoral fellow at the Ecole normale supérieure de Lyon (ENS), France (Nov. 2013–Mar. 2016). His principal axes of research are organic synthesis, polymer electrolytes, electrochemistry, and catalysis for different applications (energy storage systems and $CO_2$ valorization).

# Contents

# Preface

Sustainable development and alternative energy constituted urgent needs in the last decade. Renewable chemicals, energy and bio-resources use became challenging topics in the sustainable, renewable and green sciences. This encourages and turns primordial needs the works in certain fields as developing of new and green catalysts for chemical transformations, in the domains of energy, environmental, pharmaceutical, agro-alimentary and cosmetically applications; evaluation of bio-resources compounds largely available for many applications in energy or as additives to fuels and other applications as well as developing new synthesis routes by avoiding the use of toxic and environmentally damage materials.

This volume contains a series of six contributions as review and original papers treating 'Greenhouse Gas Conversion Efficiency in Microwave', 'Biomass Green Process' and 'Green Synthesis and Catalysis'.

Green and efficient catalyzed conversion of $CO_2$ to valuable compounds is widely developed today. Chapters 1 and 2 of the first section of this book report non-thermal plasma and microwave technologies that provide attractive and alternative route in comparison to the other classical technologies for converting greenhouse gases, such as CO2 and others, into valuable fuels and chemicals. Chapter 3, in the second section, reports catalytic conversions of biomass-derived furaldehydes towards biofuels. Chapter 4, in the third section, reports a green synthesis of oligomer calixarenes using efficient microwave heating that leads to a high selectivity and efficiency in shorter reaction time instead of the traditional thermal heating. In the last section, Chapter 5 describes an environment-friendly approach in the synthesis of catalysts based on a metal/polymeric nanocomposite particle for catalytic environmental applications. Finally, the last chapter systematically reviews the synthesis of hierarchical TS-1 zeolites and their catalytic applications in several oxidation processes at Sinopec to replace the hazardous and polluted traditional routes.

Dear readers, this volume represents a good degree of high-value research. Nevertheless, we look soon to complete the work by editing other volumes. We hope you find this book enjoyable and illuminating in similar line to our previous works *Hydrogenation-2012* and *Recent Advances in Organocatalysis*, and any comments you may have are warmly welcome.

We would like to thank all the authors for their precious contributions and their co-operation with us and in adhering to the time-right deadlines. We thank warmly the Lebanese University for giving us all the facilities to complete this volume.

Finally, we wish to express our gratitude to the staff at InTech for their kind assistance in bringing this book to fruition.

**Iyad Karamé**
Faculty of Sciences I, Lebanese University
Beirut, Lebanon

**Hassan Srour**
CPE of Lyon, CNRS-F
Lyon, France

Section 1

# Greenhouse Gas Conversion Efficiency in Microwave

# Chapter 1

# Enhancing the Greenhouse Gas Conversion Efficiency in Microwave Discharges by Power Modulation

Nikolay Britun, Guoxing Chen, Tiago Silva,
Thomas Godfroid,
Marie-Paule Delplancke-Ogletree and Rony Snyders

Additional information is available at the end of the chapter

http://dx.doi.org/10.5772/67875

**Abstract**

Scientific interest to the plasma-assisted greenhouse gas conversion continuously increases nowadays, as a part of the global Green Energy activities. Among the plasma sources suitable for conversion of $CO_2$ and other greenhouse gases, the non-equilibrium (low-temperature) discharges where the electron temperature is considerably higher than the gas temperature, represent special interest. The flowing gas discharges sustained by microwave radiation are proven to be especially suitable for molecular gas conversion due to high degree of non-equilibrium they possess. In this Chapter the optimization of $CO_2$ conversion efficiency in microwave discharges working in pulsed regime is considered. The pulsed energy delivery represents new approach for maximization of $CO_2$ conversion solely based on the discharge "fine-tuning", i. e. without the additional power expenses. In our work several discharge parameters along the gas flow direction in the discharge have been studied using various diagnostic techniques, such as optical actinometry, laser-induced fluorescence, and gas chromatography. The results show that $CO_2$ conversion efficiency can be essentially increased solely based on the plasma pulse frequency tuning. The obtained results are explained by the relation between the plasma pulse parameters and the characteristic time of the relevant energy transfer processes in the discharge.

**Keywords:** green energy, greenhouse gas decomposition, plasma-assisted $CO_2$ conversion, power modulation, optical spectroscopy, laser spectroscopy, optical diagnostics

## 1. Introduction

The basic green energy activities include reduction of the greenhouse gas emission, in particular the emission of $CO_2$, development and implementation of the green energy sources, as well as the local conversion, reforming and utilization of the greenhouse gases (see **Figure 1**). The last activities are represented by the well-known carbon capture and storage (CCS) and carbon capture and utilization (CCU) techniques being under intensive development during the last decades [1, 2]. Among the approaches suitable for local conversion of the greenhouse gases, and in particular $CO_2$, the plasma-assisted conversion based on decomposition of the molecules of interest in the low-temperature discharges is of a special interest. The efficiency of such a conversion depends on various factors, such as the proper choice of plasma source, as well as its careful optimization. The plasma source optimization in this case implies tuning the fundamental physical parameters, such as the electron density and temperature, the vibrational excitation of the molecules of interest, the discharge pressure, the residence time of the molecules in the active zone, etc. For successful optimization detailed plasma source characterization is required, preferably using nonintrusive *in-situ* plasma diagnostics methods [3] often combined with plasma modeling [4].

Since the end of 1970s the utilization of low-temperature discharges is considered as a rather promising approach for molecular gas decomposition [5, 6], which can be applied for conversion of $CO_2$ as well as the other greenhouse gases. It is well-known that the low-temperature plasma discharges represent unique media where selectivity toward the various plasma chemical processes can be realized [6, 7]. This is a result of the high level of nonequilibrium between the main

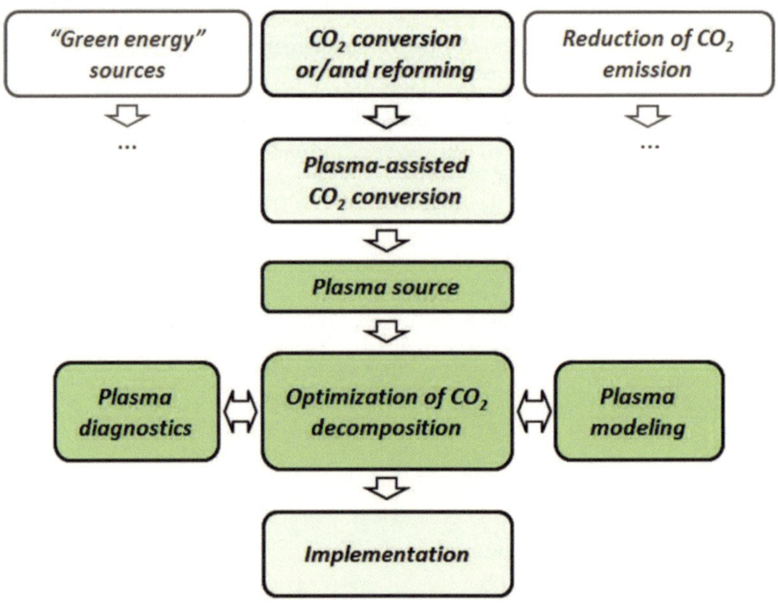


**Figure 1.** The role of plasma-assisted $CO_2$ conversion among the other green energy activities.

http://dx.doi.org/10.5772/67875

degrees of excitation in the discharge, such as electronic, translational, rotational, and vibrational ones. At the same time, as also shown previously, the vibrational excitation is particularly crucial for molecular dissociation in many cases [7, 8]. A high degree of nonequilibrium is especially pronounced in the microwave (MW) discharges where 90-95% of the applied electro-magnetic field energy is absorbed by plasma electrons further transmitting their energy to the vibrationally excited molecular states, while the gas temperature remains relatively low, typically about a thousand degree K or lower [6, 9].

A decisive role of the asymmetric vibrational mode of $CO_2$ (see below) for its efficient plasma-based decomposition has also been deduced based on the past research. The mentioned efficiency is defined, however, by several mechanisms working along with the mentioned vibrational excitation, such as: (i) fast excitation of the $CO_2$ asymmetric mode by plasma electrons as a result of the electron-vibrational (e-V) energy transfer, (ii) several orders of magnitude faster vibrational-vibrational (V-V) energy transfer, providing fast vibrational excitation of $CO_2$ molecules up to the dissociative level (≈5.5 eV), and (iii) rather slow vibrational-translational (V-T) transfer. The fast V-V process facilitates so-called vibrational "ladder climbing," i.e., fast excitation of the highly excited vibrational states of $CO_2$ result in its dissociation [6]. The low gas temperature in the discharge, on the other hand, is important for keeping the V-T energy transfer at a relatively low level, thus sustaining the vibrational excitation for a longer time. This time may vary from few μs to tens of ms, depending on the discharge parameters, as discussed below in this Chapter. As a result, the energy efficiency (defined below) up to 0.8 in the subsonic [10] and about 0.9 in a supersonic [11] gas flows may be achieved in the microwave plasmas. The nonequilibrium discharge conditions can also be effectively realized in the other discharges, for example in a gliding arc plasma (GAP) as a result of the arc length extension resulting in gas cooling [12].

Nowadays, the scientific interest to the greenhouse gas decomposition in the low-temperature discharges is growing. This is related to both low-pressure and high-pressure (atmospheric) cases. Apart from the general green energy trends, the growing interest is particularly based on the fast development of the scientific tools targeted to advanced nonintrusive discharge diagnostics as well as the powerful data processing systems, which are crucial for discharge kinetic modeling.

Talking in particular about the MW plasma sources we should note that along with their general effectiveness for molecular gas decomposition, the *power modulation effect* is known to be an additional alternative for enhancing the $CO_2$ conversion in these discharges [13]. To realize this effect, an electromagnetic wave with a filling frequency in the GHz range (serving to sustain the discharge) is periodically modulated by nearly squared pulses with a certain repetition rate, typically in the kHz range. Such modulation might be very important when the vibrational excitation is involved in a molecular decomposition process, as in the $CO_2$ case [6]. Since in the pulsed MW discharges the characteristic time of V-T energy transfer may be comparable to the typical plasma pulse repetition period (μs–ms range), a resonance-like effect between the power delivery and the energy transfer processes may be achieved. Thus, the periodic power delivery in these discharges represents an additional way for improving the efficiency of $CO_2$ conversion, at the same time requiring understanding of the corresponding physical processes. An optimization performed without increasing of the total power consumption might be extremely beneficial from the practical point of view.

In spite of the numerous works devoted to plasma-based greenhouse gas conversion in the MW discharges [11, 14, 15], dielectric barrier discharges (DBDs) [16–21], GAPs [22–24], radio-frequency (RF) discharges [25], as well as in different discharges using plasma catalysis [14, 26–28], the effects of $CO_2$ conversion and power modulation are still far from being understood fully. The number of the research works in this domain is also very limited, mainly by the theoretical speculations of the usability of pulsed plasma regime in MW and DBD cases [29], as well as by few experimental evidences of the pulsed discharge, benefits so far only shown for DBD case [17–19]. At the same time, the domain of the power modulation in MW low-temperature plasmas, representing one of the most promising nonequilibrium media for selective plasma chemistry, remains mainly unexplored. This chapter addresses the mentioned gap by demonstrating the importance of power modulation in the kHz range for improving the $CO_2$ decomposition in microwave discharges.

## 2. The basics of plasma-assisted $CO_2$ conversion

### 2.1. The structure and dissociation of $CO_2$ molecule

The structure of $CO_2$ molecule is schematically shown in **Figure 2**. This triatomic molecule possesses three vibrational modes, namely the symmetric stretch mode (with the main vibrational energy gap equal to about 0.17 eV), the double-degenerated bending mode (0.083 eV), and the asymmetric mode (0.291 eV). The different energy gaps for the listed modes define the differences in the energy transfer rates between them and the translational particle motion (V-T transfer), as discussed in Section 2.3. The vibrational excitation states for $CO_2$ are normally denoted through three vibrational quantum numbers corresponding to symmetric ($v_1$), bending ($v_2$), and asymmetric ($v_3$) vibrational modes.

One of the possible pathways for $CO_2$ decomposition is the electron impact dissociation:

$$CO_2 + e \rightarrow CO + O + e \text{ (the energy required is } \approx 5.5 \text{ eV).} \tag{1}$$

However, as a result of recombination of atomic O with vibrationally excited $CO_2$ (denoted as $CO_2^{vibr}$) another CO molecule can be produced [6], and an effective energy per one produced CO molecule becomes ≈ 2.9 eV:

$$O + CO_2^{vibr} \rightarrow CO + O_2 \text{(the energy required is } \approx 0.3 \text{ eV).} \tag{2}$$

As mentioned earlier, the actual $CO_2$ dissociation strongly relies of the e-V energy transfer resulting in the excitation of the lowest vibrational states of $CO_2$ molecule, e.g.:

$$e + CO_2 \rightarrow CO_2(0, 0, 1) + e, \tag{3}$$

where the parenthesized numbers represent the vibrational quantum numbers mentioned above. The excitation of higher vibrational states is occurring at the same time as a result of the fast energy transfer between the different vibrational states within the same vibrational mode (i.e., V-V transfer), e.g.:

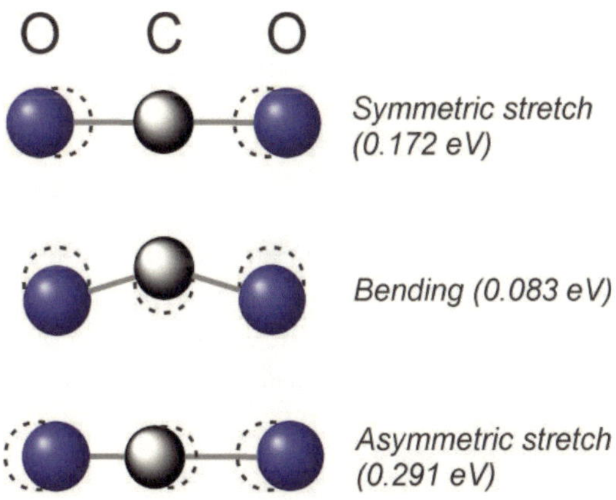


**Figure 2.** The structure of $CO_2$ molecule showing three vibrational modes. The first quantum oscillator gap energies for each mode are parenthesized.

$$CO_2(0,0,1) + CO_2(0,0,1) \rightarrow CO_2(0,0,2) + CO_2(0,0,0). \tag{4}$$

The vibrational transfer between the *different* modes (V-V′ or nonresonant transfer) is less efficient, having the typical transfer rates several times lower, according to [30].

In addition to the above-mentioned reactions, the various two- or three-body collisional processes leading to O atom recombination (some of them are exothermic) may also be important for the total energy balance in the O-containing discharges. These processes may involve both ground state ($^3$P) as well as the first excited ($^1$D) state of atomic oxygen, as well as the other states [31]. Among the typical examples are:

$$O(^1D) + O_2 \rightarrow O(^3P) + O_2, \tag{5a}$$

$$O(^3P) + O(^1D) \rightarrow O(^3P) + O(^3P), \tag{5b}$$

$$O(^3P) + O(^3P) + O_2 \rightarrow O_2 + O_2(X,0), \text{ etc.} \tag{5c}$$

### 2.2. The parameters defining $CO_2$ conversion

The $CO_2$ conversion efficiency in the discharge (as a result of $CO_2$ dissociation following by $O^-$ recombination) is usually determined as a ratio between the densities of the decomposed $CO_2$ molecules to their initial density:

$$\chi = \frac{[CO_2]_{dec.}}{[CO_2]_{init.}}, \tag{6}$$

where $[CO_2]_{dec.}$ and $[CO_2]_{init.}$ refer to the decomposed and initial $CO_2$ molecular density, respectively. Let us note that in the pure $CO_2$ case, $\chi$ can be defined simply as $[CO]/[CO_2]_{init.}$, where [CO] is the density of CO molecules produced.

At the same time, the energy efficiency $\eta$ is normally defined via the enthalpy of $CO_2$ dissociation ($\Delta H_{CO_2} = 2.9$ eV) and the energy spent for production of one CO molecule ($E_{CO}$) [6]:

$$\eta = \frac{\Delta H_{CO_2}}{E_{CO}}. \tag{7}$$

In a general case, both $\chi$ and $\eta$ quantities are defined locally, as the supplied energy might be different depending on the point of interest in the discharge. Combining last two expressions and defining the specific energy input (SEI) as the energy delivered per a single $CO_2$ molecule in a certain discharge volume (usually expressed in eV or eV/molec), the energy efficiency yields:

$$\eta = \chi \frac{\Delta H_{CO_2}}{\text{SEI}}. \tag{8}$$

The specific energy can be usually determined via the power $P$ applied to the discharge and the flux of the gas $F$ in the discharge tube:

$$\text{SEI} = \frac{P}{F}, \text{ or using more convenient units, } \text{SEI(eV/molec)} = 0.014\, \frac{P(\text{W})}{F(\text{slm})}, \tag{9}$$

where $P$ is in Watts and $F$ is in slms (slm stands for standard liter per minute).

### 2.3. The main energy transfer channels in $CO_2$ plasma

In molecular plasma, the interaction between translational, rotational, and vibrational degrees of freedom, involving also the plasma electrons, leads to formation of the numerous energy exchange (relaxation) channels responsible for excitation or depletion of the corresponding energy subsystems. Some of these channels are especially important for understanding of the plasma-assisted $CO_2$ decomposition, due to the crucial role of the vibrational excitation in this process. Among them are the translational-translational (T-T), rotational-translational (R-T), as well as the e-V, V-V, and V-T channels mentioned above. The physical nature as well as the corresponding characteristic times for these energy relaxation mechanisms are described below.

#### *2.3.1. T-T and R-T energy transfer*

The T-T energy transfer is responsible for gas thermalization (i.e., establishing kinetic gas temperature). The characteristic time of this process can be estimated via the mean free path of the gas atoms and molecules and their mean velocity, being typically in the μs range for the gas pressure of several Torr and gas temperature of several hundred degree K. The R-T process, on the other hand, shows how fast the rotational degrees of freedom in molecular gas will be in the equilibrium with the translational (kinetic) motion of the gas particles. In order to estimate the characteristic time of this process, the model proposed by Parker [32] can be used (see Ref. [6] for further details). According to this model, the characteristic time of the R-T process can be determined based on the number of collisions necessary for equilibrium between the rotational and translational degrees of freedom in discharge. The estimation of this time (for variety of gases) at about 20 Torr of the gas pressure leads to the values $\tau_{R-T} < 0.1$ μs, so the R-T equilibrium can be considered nearly instantaneous in the time scale related to this work, as compared below.

### 2.3.2. *e-V energy transfer*

For the efficient plasma-based molecular decomposition in the most low-temperature discharge cases, an efficient transfer of the electron energy to the vibrational degrees of freedom of the corresponding molecules is critical. In the case of $CO_2$, the characteristic time of this process can be determined through the corresponding rate coefficient $k_{e-V}$ as

$$\tau_{e-V} \sim (k_{e-V} \cdot n_e)^{-1}, \tag{10}$$

where $n_e$ is the electron density in plasma. Assuming that the electron density in the microwave discharge in the considered gas pressure range is about $10^{12}$–$10^{13}$ cm$^{-3}$ [33], and taking $k_{e-V} \approx 10^{-8}$ cm$^{-3}$/s [34] corresponding to excitation of the lowest $CO_2$ vibrational level, we obtain $\tau_{e-V}$ ~10–100 μs. As we can see, under our conditions this process is much slower than the R-T energy relaxation.

### 2.3.3. *V-V energy transfer*

The rate of establishing vibrational equilibrium within each vibrational mode of a certain molecule is defined by the V-V energy transfer and the corresponding characteristic time. In the case of $CO_2$ molecule, the V-V rates are roughly comparable for all three vibrational modes (see **Figure 2**) having the differences within the order of magnitude [30]. A rough estimation of the V-V characteristic time for the asymmetric mode of $CO_2$ gives

$$\tau_{V-V} = (k_{V-V} \cdot [CO_2])^{-1} \sim 3 \text{ ns, (at 20 Torr and 300K)}, \tag{11}$$

where $k_{V-V}$ is the corresponding rate coefficient ($k_{V-V} \sim 5 \cdot \times 10^{-10}$ cm$^{-3}$/s for the first energy levels of the asymmetric mode, according to Ref. [6]), and $[CO_2]$ is the ground state number density of $CO_2$ molecules in the discharge estimated in our case using the actual values of the gas pressure and temperature. As we can see that the V-V process is nearly instantaneous being faster than both R-T and e-V processes.

### 2.3.4. *V-T energy transfer*

While the plasma electrons transfer their energy to the $CO_2$ vibrational modes, the vibrational excitation might be suppressed by the translational motion of gas particles, as a result of V-T energy transfer. This effect should be considered harmful for the efficient $CO_2$ decomposition, taking into account the importance of vibrational excitation in this case. The characteristic time of the V-T process, $\tau_{V-T}$, varies significantly spending on the gas pressure and gas temperature and can be estimated based on several models available elsewhere [5, 6]. One of them is represented by a semiempirical expression proposed by Millikan and White [35]:

$$\tau_{V-T} = \frac{1}{p}\exp\left[1.16\cdot10^{-3}\mu^{\frac{1}{2}}(\hbar\omega)^{\frac{4}{3}}\left(T_0^{-\frac{1}{3}} - 0.015\mu^{1/4}\right) - 18.42\right], \tag{12}$$

where $\tau_{V-T}$ is the relaxation time (s), $p$ is the gas pressure (atm), $T_0$ is the gas temperature (K), $\mu$ is the reduced mass of the colliding molecules (a.m.u.), and $\hbar\omega$ is the vibrational gap for the corresponding vibrational mode (also expressed in K). The expression (12) is very sensitive to the gas temperature as well as to the vibration mode under consideration, giving extremely long characteristic times for the $CO_2$ asymmetric mode (~$10^5$ s). This might be related to the fact that this expression describes well only the first (lowest) vibrational states of the $CO_2$ bending mode, for which (at 20 Torr and 1000 K) the estimates give $\tau_{V-T} \approx 15$ μs, whereas at 300 K this value increases to about 150 μs.

The role of the gas temperature in the V-T energy transfer is additionally illustrated in **Figure 3**, where the $\tau_{V-T}$ time is calculated in the 300–3000 K temperature range. The calculations clearly show that $\tau_{V-T}$ can vary by nearly two orders of magnitude in the mentioned temperature

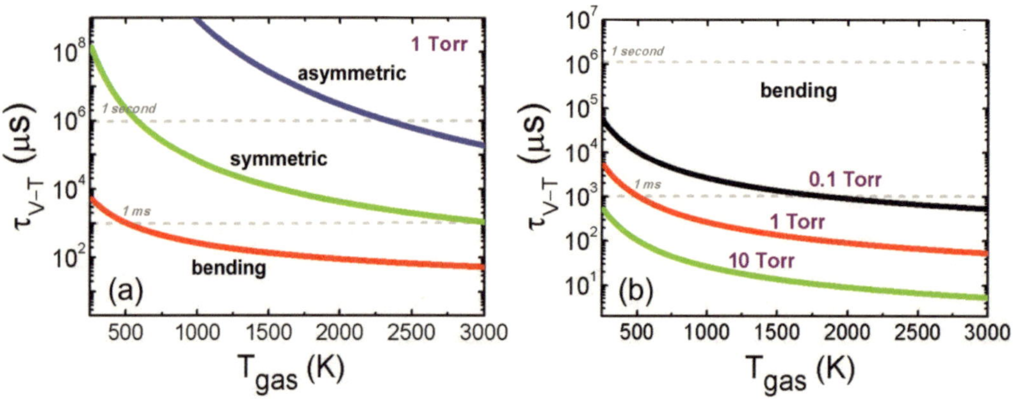


**Figure 3.** The characteristic time of V-T relaxation as a function of gas temperature calculated (based on Ref. [35]) for three vibrational modes of $CO_2$ at 1 Torr (a), and for the bending vibrational mode at different gas pressure (b).

| Process | Characteristic time (at 20 Torr) | Comment | Source |
|---|---|---|---|
| T-T | <0.1 μs | $T_{gas} = 300$ K | [36] |
| R-T | <0.1 μs | $T_{gas} = 300$ K | [6, 32] ** |
| e-V * | ~10 μs | $T_e = 1.5$ eV, $n_e \sim 10^{13}$ cm$^{-3}$ | [34, 37] ** |
| V-V | <0.01 μs | $T_{gas} = 300$ K | [6] |
| V-T (bending) | 150 μs | $T_{gas} = 300$ K | [6, 35] ** |
| V-T (bending) | 15 μs | $T_{gas} = 1000$ K | [6, 35] ** |

* Done for the symmetric stretch and bending $CO_2$ vibration modes: (1,0,0) and (0,1,0).
** In case of multiple literature sources the results are averaged.

**Table 1.** The estimates for the characteristic time of the main energy transfer processes described in the Section 2.3.

http://dx.doi.org/10.5772/67875

range. The calculations also point out on a primary role of the $CO_2$ bending mode in this process, as the one corresponding to the fastest V-T transfer (see **Figure 3(a)**). Naturally, the $\tau_{V-T}$ time is inversely proportional to the gas pressure since the number of collisions per unit time defining the V-T energy exchange rate is directly proportional to pressure (see **Figure 3(b)**).

The rough estimates for the characteristic times corresponding to the main energy exchange processes described in this section are summarized in **Table 1**.

## 3. The experimental

### 3.1. The plasma sources used

The pulsed microwave discharges (surfaguide-type) have been used as the plasma sources in this study. In these discharges, plasma is sustained by an electromagnetic wave with the filling frequency in the microwave range (either 0.915 or 2.45 GHz in our case) coming out of two orifices in the surfaguide [38], as schematically shown in **Figure 4**. In our case, the electromagnetic radiation has been modulated by the nearly square pulses with the repetition frequency ranging from 0.5 to 30 kHz. The duty ratio of the pulses was kept equal to 50%. The discharges were sustained in the quartz tubes (14 mm in diameter and 31 cm long) in which the gas flow has been regulated by digital mass flow controllers. Each quartz tube was additionally cooled by a flow of Si oil (~2 l per minute) having the temperature of about 5°C (in the 0.915-GHz system) or 10°C (in the 2.45-GHz system). The total gas flow rate has been varied in the range from about 0.08–2.7 slm. Both pure $CO_2$ and $CO_2 + 5\%\ N_2$ gas mixtures have been utilized. The time-averaged power applied to the discharge was always fixed at the level of either 0.4 kW (2.45-GHz system) or 1.0 kW (0.915-GHz system). The reflected electromagnetic radiation has been minimized in each plasma source using three-stub automatic tuning systems.

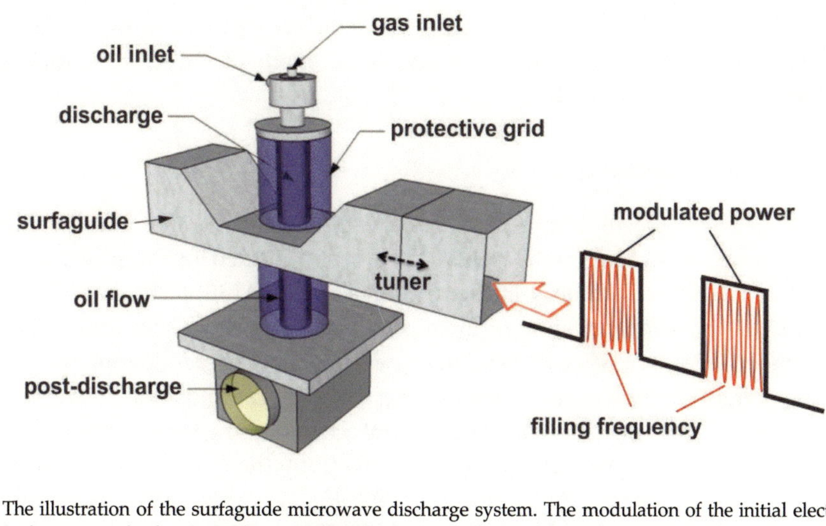


**Figure 4.** The illustration of the surfaguide microwave discharge system. The modulation of the initial electromagnetic wave (filling frequency of either 2.45 GHz or 0.915 GHz) by the kHz pulses is shown schematically.

The reflected power was always around 5% for 2.45-GHz system, and totally negligible (presumably <1%) in the case of 0.915-GHz system. Further experimental details related to the mentioned microwave systems can be found elsewhere [15, 39]. The diagnostics of the microwave discharge has been undertaken both in the discharge active zone (i.e., near the waveguide excitation point) and in the postdischarge (at about 40 cm below the plasma excitation point), as described in the following section.

### 3.2. The diagnostic techniques used

#### *3.2.1. Diagnostics in the discharge zone*

Optical emission spectroscopy (OES), including emission actinometry and ro-vibrational analysis, has been applied for characterization of the $CO_2$ conversion efficiency as well as the gas temperature in the discharge zone. Gas temperature has been also monitored by a thermocouple at the beginning of the postdischarge (i.e., about 17 cm below the excitation point). The rotational band from the CO Angstrom system corresponding to the $B^1\Sigma^+$ ($v' = 0$) – A $^1\Pi$ ($v'' = 1$) optical transition has been used for the rotational temperature determination in the discharge. Gas temperature has been assumed equal to rotational temperature of CO molecules, based on the analysis undertaken elsewhere [15, 40].

In order to determine the $CO_2$ conversion efficiency in the discharge area, the optical actinometry method based on the addition of a small amount of molecular nitrogen (5% in our case) to the $CO_2$ gas has been applied. This method is based on the measurement of the emission lines ratio between the unknown species in the discharge (CO) and the known admixture ($N_2$). As a result, assuming corona excitation in the discharge volume, the absolute density of species of interest can be determined, as performed recently by Silva et al. [15]. The relative error of this method is supposed to be <10%, based on our estimations.

An Andor Shamrock-750 monochromator having 0.75 m of focal length and equipped with an Andor iStar-740 series intensified charge coupled device (ICCD) camera has been used for spectral acquisition in this work. The spectral resolution during the measurements was equal to about 0.05 nm (at 500 nm). The accumulative ICCD mode of the spectral acquisition has been used for all the measurements.

#### *3.2.2. Diagnostics in the postdischarge zone*

In the case of 2.45-GHz plasma source, the products of $CO_2$ dissociation (such as CO ground state molecules and O ground state atoms) have been detected in the postdischarge area using a two-photon absorption laser-induced fluorescence (TALIF) technique [41]. This technique is based on the laser excitation of the molecular or atomic species in the discharge or postdischarge by a simultaneous absorption of two laser photons, following by a spontaneous emission of light (fluorescence) corresponding to an optical transition between the upper (excited) state and the intermediate state. The spectral schemes using laser excitation at 225.6 nm (for O atom) or 230.07 nm (for CO molecule) following by the fluorescence at 844.7 nm (O) or 483.5 nm (CO) have been applied in this work [42, 43]. The summary of the corresponding spectral transitions used for O and CO detection by TALIF technique is given in **Table 2.**

| Spectral parameter | Values | |
|---|---|---|
| **Specie of interest** | **O [42]** | **CO [43]** |
| Lower state | $O(^3P_2)$ | $CO(X^1\Sigma^+)$ |
| Upper (laser-excited) state | $O(^5P)$ | $CO(B^1\Sigma^+)$ |
| Energy gap | 10.74 eV | 10.78 eV |
| Laser excitation wavelength | 225.6 nm | 230.07 nm |
| Fluorescence wavelength | 844.68 nm | 483.50 nm |
| Bandpass filter used | 840 nm | 480 nm |

**Table 2.** The spectral transitions for TALIF diagnostics of the ground state O and CO in the postdischarge of the 2.45 GHz MW source used in this work.

A Sirah dye laser working at 10 Hz of repetition rate and having 5 ns of the pulse duration pumped by a Spectra Physics YAG:Nd laser has been utilized for TALIF diagnostics. A Coumarin 450 dye solution (in ethanol) has been used in the dye laser. During the measurements, the laser pulses were not synchronized with the plasma pulses, thus giving the time-averaged values of the corresponding ground state densities in the postdischarge. The provided averaged values are supposed to reflect an overall system performance regarding $CO_2$ conversion, being especially interesting from the applications point of view.

Another type of diagnostics applied to the 0.915-GHz MW discharge was a gas chromatography (GC) technique. The GC technique is based on the different gas elution time on the analyzer walls, representing *ex-situ* time-averaged gas analysis, which has been used for characterization of various gas mixtures, including the products of the $CO_2$ decomposition [11]. In spite of being an *ex-situ* technique, GC provides the results that can be compared with the laser-based (*in-situ*) techniques for the stable dissociation products, which is valid for CO ground state molecules. In our case, a Bruker 450-GC gas chromatograph equipped with a sampling system has been used for the postdischarge characterization of the dissociation products in the 0.915-GHz MW source. In the described GC system, a low-pressure gas sample is diluted with carrier gas (argon) before its injection into the gas chromatograph for the further *ex-situ* analysis.

## 4. Optimization of $CO_2$ conversion in microwave plasma

### 4.1. The emission spectroscopy analysis

$CO_2$ decomposition in a flowing gas discharge can often be noticed visually by changing the color of the discharge before and after the excitation point (waveguide position) in the discharge tube. This corresponds to the formation of CO molecules in the so-called discharge "active zone" (being approximately 6 cm wide in our case, according to Ref. [15]), where the decomposition process is mainly taking place. In the case of CO, the observed emission corresponds to the CO Angstrom band ($CO(B^1\Sigma^+)$ – $CO(A\ ^1\Pi)$) and partially to the third positive (3P) CO band ($CO(b^3\Sigma^+)$–$CO(a\ ^3\Pi)$) [44], as shown in **Figure 5**. In the $CO_2$-$N_2$ gas mixture, these effects are qualitatively similar to the pure $CO_2$ case. The O atom emission triplet around 777 nm is also clear in both cases, whereas the N atom emission around 821 nm is rather negligible in the case of $N_2$ admixture, as shown in **Figure 5(b)**.

The emission spectra corresponding to the $CO_2$-5%$N_2$ gas mixture and taken in a wider spectral range are presented in **Figure 6**. As we can see that at higher gas pressure, a much stronger contribution of the $N_2$ and $N_2^+$ molecular bands such as $N_2$ second positive band ($N_2(C\ ^3\Pi_u)$-$N_2(B\ ^3\Pi_g)$), $N_2^+$ first negative band ($N_2^+(B^2\Sigma_u^+)$-$N_2^+(X^2\Sigma_g^+)$), and $N_2$ first positive band ($N_2(B\ ^3\Pi_g)$-$N_2(A^3\Sigma_u^+)$) is evident. The contribution of the CO Angstrom ro-vibrational band is rather strong in both cases. The structure of all the observed CO rotational

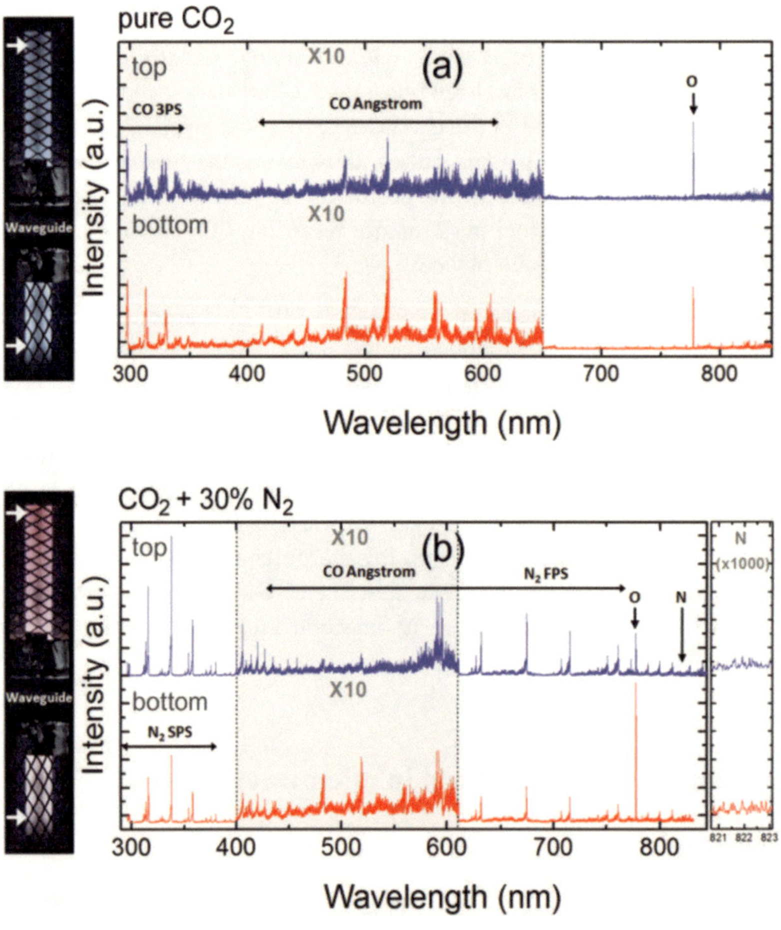


**Figure 5.** Time-averaged emission spectra measured in the 2.45-GHz MW discharge at two different special positions for pure $CO_2$ (a) and $CO_2$ + 30%$N_2$ mixture (b). Gas pressure is about 2 Torr. Bold arrows indicate the top and bottom measurement points. The spectral correction is applied in both cases. The corresponding discharge photographs are shown on the left side.

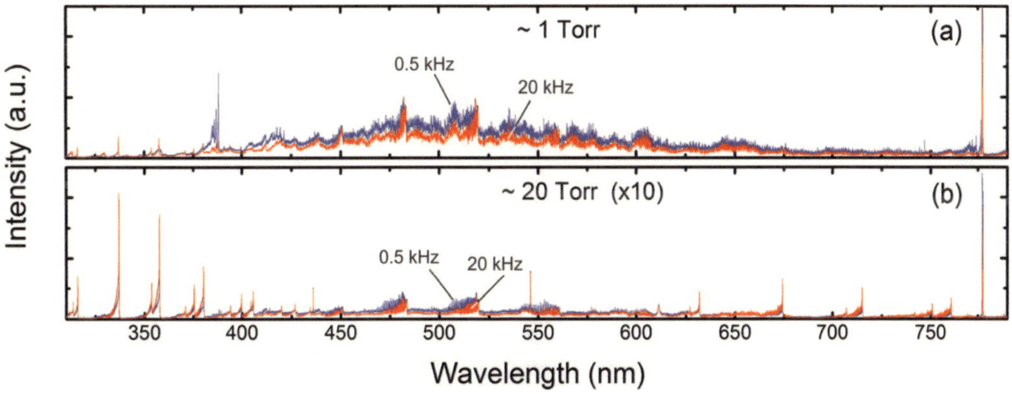


**Figure 6.** Time-averaged emission spectra taken in the 2.45-GHz MW discharge at high and low plasma pulse frequency at 1 Torr (a) and ~20 Torr (b) in the $CO_2 + 5\%N_2$ gas mixture. No spectral correction is applied.

bands is different at high and low pulse repetition frequency. At low frequency, they are more elongated toward the shorted wavelengths corresponding to higher gas temperature in the discharge, confirmed by calculations, whereas the gas temperature is essentially lower at high frequency.

In addition, the low pressure spectra possess much more pronounced continuum band in the 400–600 nm range (**Figure 6(a)**). Such a strong contribution of the continuum band is likely related to the chemiluminescence induced by the CO–O recombination process, as also detected in Ref. [15] and analyzed in Refs. [45, 46].

### 4.2. $CO_2$ conversion efficiency and related results

The main results on the $CO_2$ conversion efficiency ($\chi$) and energy efficiency ($\eta$) are described in this section aiming at comparison of two mentioned MW plasma sources. The obtained data are presented as a function of the plasma pulse repetition frequency (f) aiming at the clarification of namely the effect of plasma power modulation on the $CO_2$ conversion.

The relative density of the CO ground state molecules detected by the TALIF technique in the postdischarge of the considered MW plasma sources is shown in **Figure 7**. The beneficial effect of power modulation is evident in this case leading to a fourfold increase in the CO density (so the corresponding $CO_2$ conversion efficiency) at low pulse frequency. The maxima of $\chi$ are observed at about 0.5 kHz (for the 2.45-GHz system) and at about 0.8 kHz (for the 0.915-GHz system). Apart from the different positions of these maxima, in the 2.45-GHz case, maximum appears to be much narrower than that detected in the 0.915-GHz discharge case.

As also clear from **Figure 7(a)**, the O production is strongly suppressed at low pulse frequencies (below 1 kHz), when the dissociation of $CO_2$ reaches its maximum. At the same time, the $O_2$ density also has a maximum at low frequency (detected in the 0.915-GHz system though), pointing out on the efficient O-recombination process under these conditions (see Figure 7(b)). Note that the CO and $O_2$ densities determined by GC are not different by a factor of two in this

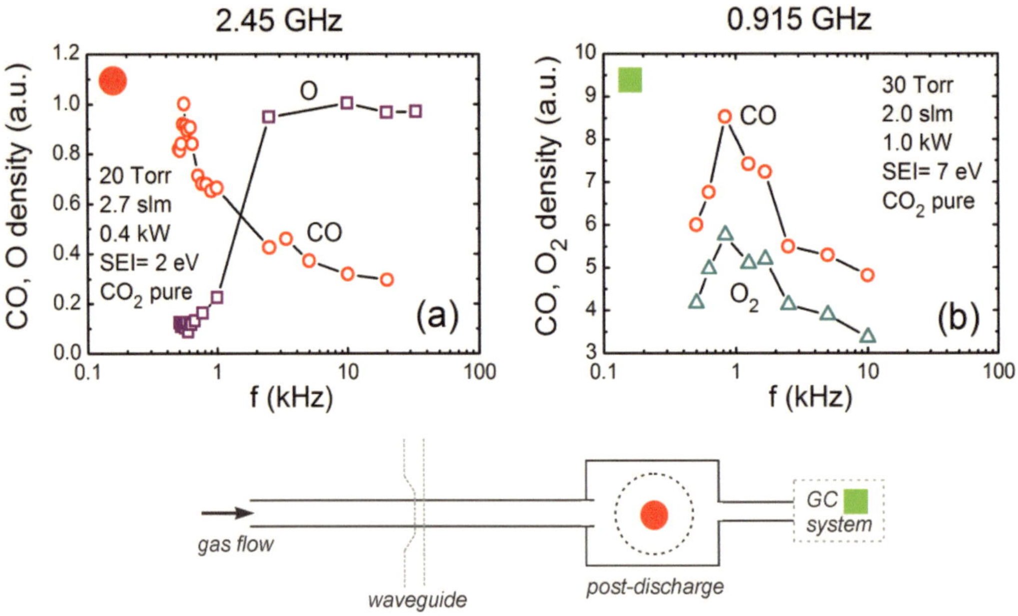


**Figure 7.** (a) Normalized density of CO and O in the postdischarge as a function of the plasma pulse repetition frequency (f) measured by the TALIF technique in the 2.45-GHz MW system. (b) The relative density of CO and $O_2$ molecules measured in the 0.915-GHz MW system by GC. The points of measurement are indicated below.

case, due to the fact that calibration of our GC system response was not performed. Finally, the CO density decay slopes measured in two different plasma sources reveal rather similar behaviors.

The data on the $CO_2$ conversion efficiency as a function of the plasma pulse frequency determined by optical actinometry in the 2.45-GHz system (in the discharge area) and by GC in the 0.915-GHz system (in the postdischarge area) are compared in **Figure 8**. As we can see that there is a fourfold difference between the observed maximum values of $\chi$ in these two cases. This is related to the power differences between the considered plasma sources as well as to the fact that the actinometry measurements were performed in the center of the discharge tube where the $CO_2$ conversion is not yet fully accomplished. After the corresponding corrections, the obtained conversion efficiency values appear to be very similar for both systems. The energy efficiency values corresponding to the observed $\chi$ maxima in this case are 0.14 and 0.16, respectively. In addition, in the case of **Figure 8(a)**, the $CO_2$ conversion curve does not reveal a clear maximum at low plasma pulse frequency, which is only present in the 0.915-GHz case (**Figure 8(b)**). This phenomenon might be related to the gas displacements in the discharge tube, as well as to the differences in the discharge geometry. The physical reasons for the observed behavior of $\chi$ and CO production in the postdischarge are discussed in the following section.

An interesting behavior of the gas temperature in the discharge tube along the gas flow direction has also been detected, as a result of combination of ro-vibrational spectral analysis and

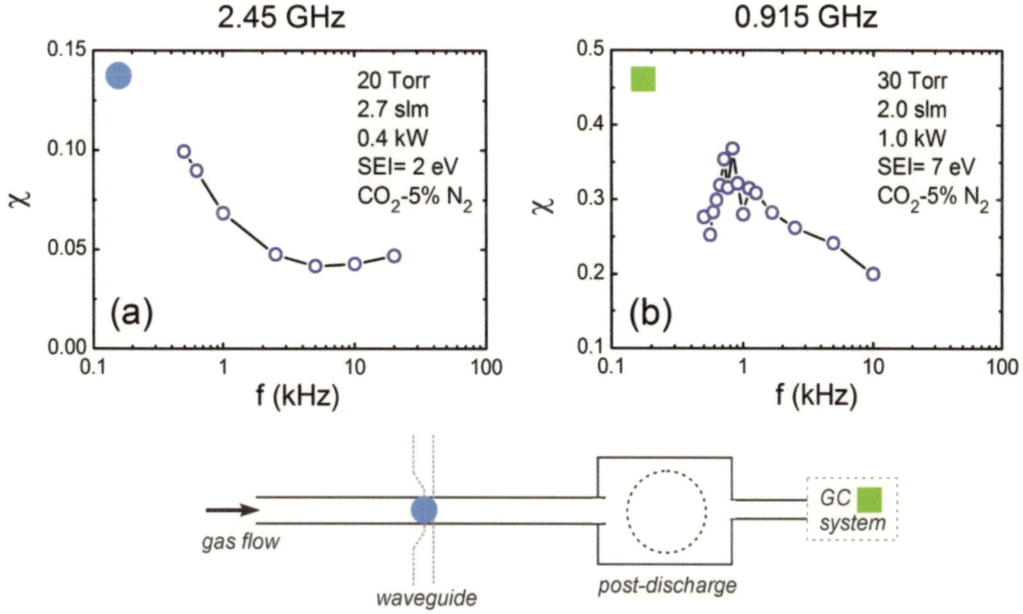


**Figure 8.** The $CO_2$ conversion efficiency $\chi$ as a function of the plasma pulse repetition frequency (f) measured in the discharge region of the 2.45-GHz MW system by optical actinometry (a), and in the postdischarge region of the 0.915-GHz MW system by GC (b). The energy efficiency corresponding to the maximum $\chi$ values is equal to 0.14 (a) and 0.16 (b). The points of measurement are indicated below.

thermocouple measurements. The corresponding results are shown in **Figure 9**. The gas temperature in the discharge zone has been measured in this case using the rotational band of CO, following by the Boltzmann plot approximation for the obtained rotational populations, as described elsewhere [15, 47]. As a result, the error bars in **Figure 9(a)** correspond to the error of Boltzmann fit applied to the rotational distributions for each data point.

As we can see that a trend for $CO_2$ conversion efficiency shown in **Figure 8(a)** clearly correlates with the one obtained for the gas temperature in the discharge zone, shown in **Figure 9(a)** for the 2.45-GHz MW source (the point of measurements is indicated by a dot). At the same time, at the end of the discharge tube (indicated by a square), the temperature behavior is roughly opposite. In this case, the gas temperature is somewhat reduced at low plasma pulse frequencies (see **Figures 9(b)** and **9(c)**). The observed temperature reduction is especially clear in the case of 0.915-GHz plasma source, when the gas temperature drops by nearly 200 K at low pulse frequency. In this case, we can talk about the existence of a temperature gradient established between the excitation point in the discharge and the end of the discharge tube (beginning of the postdischarge). Apparently, this gradient is much larger at low pulse repetition frequencies (about 600 K in our case, based on **Figure 9(a)** and **9(b)**) compared to the high frequency (only about 300 K). Based on the data shown in **Figure 9(c)**, one can speculate that this effect might be also similar in the 0.915-GHz plasma source case (for which the $T_{gas}$ data in the discharge area are not available).

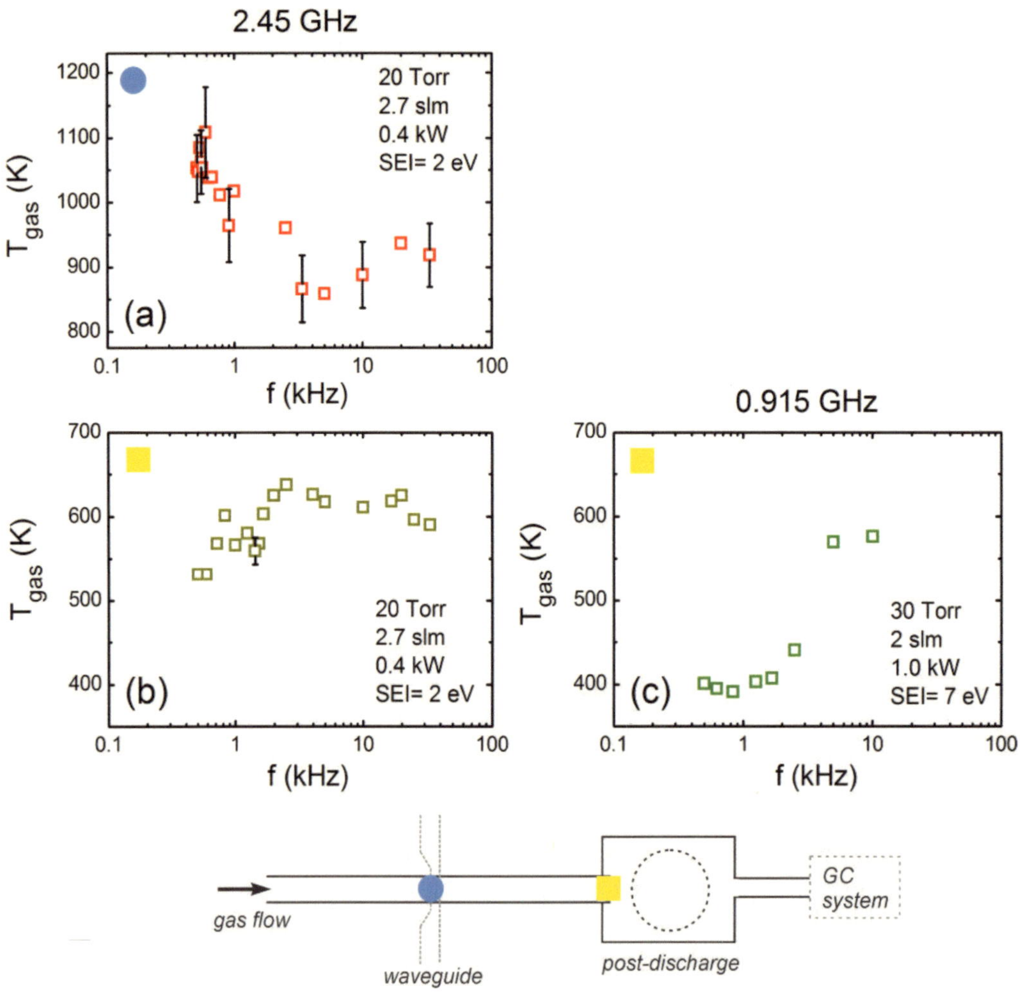


**Figure 9.** The evolution of the gas temperature as a function of plasma pulse repetition frequency (f) measured in the discharge area (a) and at the end of the discharge tube (b) in the 2.45-GHz MW system, as well as at the end of the discharge tube in the 0.915-GHz MW system (c). Pure $CO_2$ gas is used. The points of measurement are indicated below.

### 4.3. Discussion

Several physical effects should be taken into account for proper explanation of the observed $CO_2$ conversion efficiency as a function of the plasma pulse frequency, as well as its relation to the measured gas temperature, both in the discharge and the postdischarge areas. Among the main physical phenomena responsible for the efficient molecular decomposition, the vibrational excitation of $CO_2$ molecules, the gas displacement in the tube, as well as V-V and V-T relaxation processes in the discharge should be considered. The CO rotational temperature, at the same time, being determined using a CO rotational emission band, is supposed to be an indicator for the other important processes in the discharge, such as V-T energy transfer. This temperature is supposed to be in equilibrium with the kinetic gas temperature rather quickly (μs scale) as a result of the fast R-T transfer, according to our estimations (see **Table 1**), giving nearly instantaneous image of the gas heating.

According to our estimations, the plasma electrons transfer their energy to the (lowest) $CO_2$ vibrational states during the time $\tau_{e-V} \sim$ 10–100 μs. This time might be somewhat smaller at higher gas pressures, as one may expect a slight increase in $n_e$ in this case [33]. At the same time, the equilibration of the vibrational distributions for each vibrational mode is taking place nearly instantaneously, with a time constant in the order of magnitude of 0.01 μs or less, estimated for our discharge conditions. Finally, the loss of the vibrational excitation as a result of vibrational energy transfer to the translational gas motion through the bending vibrational mode (resulting in gas heating) takes place with a characteristic times ranging roughly between 15 and 150 μs in our case, depending on the gas temperature. Let us also note that the V-T energy exchange occurs faster in the discharge active zone, where the gas temperature may exceed 1000 K (thus $\tau_{V-T} <$ 15 μs), and slower in the other parts of the discharge when the temperature might be close to the room temperature ($\tau_{V-T} \sim$ 150 μs).

The gas velocity (i.e., gas displacement) in the tube is another key parameter defining the residence time for $CO_2$ molecules in the discharge active zone (~6 cm wide in our case). The rough estimations for the gas velocity based on the ideal gas low give the values of about 40 m/s in the 2.45-GHz source case (at 20 Torr, 1100 K, and 2.7 slm of a total gas flow), and about 30 m/s in the 0.915-GHz case (at 30 Torr, 1600 K [48], and 2 slm of a total gas flow).

Taking into account the above-mentioned estimations, the observed processes along the gas flow direction can be explained as follow. At first, in the discharge zone (the excitation point), the $CO_2$ vibrational excitation is initiated by the fast e-V energy transfer. The e-T and T-T energy exchange channels, at the same time, may contribute to the overall gas heating. In addition, the gas heating due to the V-T transfer should also occur, which is supposed to be more efficient at longer plasma off-times (i.e., at low frequencies). Even though the V-T characteristic time is rather short according to our estimations (~15 μs at 1000 K), its contribution at lower pulse repetition frequency is supposed to be more pronounced, likely resulting in the increase of gas temperature at low frequency measured in our case (see **Figure 9(a)**). Note that the gas temperature has been determined in this work based on the time-average spectral data and the time-resolved measurements may be necessary for a full clarification of the temperature evolution. The additional experiments [49] indicate that the vibrational temperature of the $N_2(X)$ ground state molecules in the active zone decays rapidly under the increase of the pulse repetition frequency, following the trends obtained for CO density (**Figure 7(a)**) and $CO_2$ conversion (**Figure 8(a)**). Rather fast nonresonant V-V′ energy exchange taking place between $N_2$ and $CO_2$ (see Ref. [50] and refs therein) leads to similar expectations for the $CO_2$ vibrational temperature as well. This leads to a conclusion that vibrational excitation is the main reason for higher $CO_2$ dissociation at low pulse frequency, in spite of the fact that the V-T transfer is also enhanced in this case (gas temperature increase is observed). A gradual decay in the $CO_2$ dissociation (**Figure 8**) leading to a weaker CO production (**Figure 7**) at high frequencies is most probably related to a less efficient e-V transfer at shorter plasma pulse durations. The top estimations of the e-V transfer time in our case correspond to the pulse frequency of about 5 kHz, only roughly correlating with the obtained data still requiring final clarification first of all based on the precise measurements of the electron density in the discharge active zone.

The presence of maximum in the frequency dependences of $\chi$ (0.915-GHz case, **Figure 8**) and CO density in the postdischarge (**Figures 7** and **8(b)**) at low pulse frequencies deserves special attention, since the gas displacement in the tube may play a decisive role in this effect. Three cases can be considered explaining the observed data, namely: (i) a slow gas displacement, when the gas displacement is small between two consecutive plasma pulses, (ii) a "resonant" gas displacement, when the gas displacement time in the active zone is almost equal to the plasma pulse duration, and (iii) a fast gas displacement, when the gas displacement time is shorter comparing to the time between two plasma pulses. Since in our case, the gas velocities are comparable for both the 2.45-GHz (~40 m/s) and 0.915-GHz (~30 m/s) systems, it can be shown that nearly the resonant case is realized at low plasma pulse frequency (0.5 kHz) in both systems. This results in a nearly maximum system performance in terms of $CO_2$ decomposition and energy efficiency. More pronounced $\chi$-maximum found in the 0.915-GHz case, as well as its position shifted toward higher plasma pulse frequency values may be a result of the differences in system geometry as well as the errors related to gas pressure and especially gas temperature determination, which are the critical parameters for gas velocity.

We should note that at lower pulse frequencies (or higher gas flows) one may expect a significant drop in the $CO_2$ decomposition, as the fast gas displacement limit will be achieved and some portions of the passing gas will remain untreated by plasma. On the other hand, at higher pulse frequencies, as observed in our case, a considerable drop in the $CO_2$ conversion should be likely explained by a combination of several factors, such as (i) a decrease of the e-V transfer contribution at shorter pulse durations, (ii) a decrease of the role of dissociative recombination of $CO_2^+$ (via the reaction: $e + CO_2^+ \rightarrow CO + O$, see Ref. [29]) in this frequency range, as suggested by Silva et al. [49], (iii) decomposition of CO molecules in the active zone when residence time is too long. The third argument, however, is supposed to play a minor role, due to the synchronous changes of both CO and $O_2$ densities observed in the postdischarge detected by GC, as shown in **Figure 7(b)**.

Finally, some attention should also be given to the gas temperature differences in the excitation point and at the beginning of the postdischarge, as well as to the corresponding temperature gradient between these points. At low plasma pulse frequency, the high values of both gas temperature and $CO_2$ conversion are observed, whereas the O ground state density reaches its minimum. The gas temperature at the beginning of the postdischarge is rather low in this case, resulting in a high temperature gradient between the tube center and its end. At high frequency, on the other hand, $CO_2$ conversion drops several times, along with the CO density, measured in the postdischarge. The O ground state density in the postdischarge is roughly 10 times higher in this case (see **Figure 7(a)**). Also, the gas temperature is getting lower in the discharge area and higher in the postdischarge (comparing to the low frequency case), flattening the mentioned temperature gradient.

Based on our experimental data, the observed gas temperature phenomena may be explained by the O atom recombination. Considering two main ways of this recombination, namely the reaction (2) and reactions (5) mentioned above, we can conclude that at low frequency ground state, O gets efficiently recombined either with $CO_2^{vibr.}$ or with atomic/molecular oxygen. The additional heat released as a result of (some) recombination processes (along with the e-T energy transfer) is a probable reason for the gas heating in the discharge area. As a result of the efficient O-recombination, the delivery of ground state O to the postdischarge is significantly reduced in this case, the corresponding heat release is reduced as well, resulting in a low temperature. On the other hand, at high pulse frequencies CO production drops (partially due to the reduced O-$CO_2^{vibr}$ recombination), letting more O ground state atoms to be formed as a result of electron impact dissociation in the discharge area and to be transported to the postdischarge. The O-recombination processes, other than O-$CO_2^{vibr}$ one, are still taking place all the way down to the postdischarge, thus "blurring" the hear release along the discharge tube and flattening the temperature gradient between the excitation point and the end of the tube. The O-O recombination on the termocouple surface might also be an important factor contributing to the observed tempearture elevation in this case.

The given explanation, however, describes the observed temperature behavior only in the first approximation, do not taking into account numerous additional O-recombination reactions, e.g., those involving excited O states (such as $O(^1D)$ state), as well as the processes in which O ions are involved (such as $O^+ + O_2(a) \rightarrow O_2^+ + O(^3P)$ or $O^+ + O^- \rightarrow O(^3P) + O(^3P)$, see Ref. [31]), which may additionally contribute to the O ground state density distribution along the discharge tube, as a result of quenching of the excited O states. The reactions (5a) and (5b) mentioned above also correspond to this case. This described temperature effects may still need a further clarifications in the future, based on the kinetic discharge modeling. It is already clear, however, that these effects may play a key role for the future optimization of the plasma-based $CO_2$ conversion in the microwave gas-flowing discharges, since by lowering the gas temperature the lifetime of $CO_2$ vibrational excitation might be significantly enhanced, which is favorable for efficient $CO_2$ conversion.

### 4.4. Comparing with literature

In this section, the most prominent results obtained based on the power modulation in the pulsed microwave plasma considered in this work are compared with the available literature data. The most competitive literature results have been chosen for this purpose, representing microwave, DBD, and gliding arc discharges. The discharges operating with catalysis, i.e., using a plasma-catalyst synergy, are not considered (except for one example). The corresponding data are summarized in **Figure 10**.

As we can see, the most competitive results are grouping around the diagonal line corresponding to the value of SEI of 2.9 eV/molec, as the virtual limit of the $CO_2$ conversion (when $\chi = \eta = 1$) can be reached only in this case. The beneficial effect of the plasma power modulation (indicated by solid arrows) resulting in about twofold increase of the $CO_2$ conversion and energy efficiencies is evident for both microwave plasma sources described in this chapter. In addition, the effect of plasma catalysis studied by Chen et al. [28] in the 0.915-GHz source is also given for illustration (indicated by dashed arrow). More than a twofold gain for both $\chi$ and $\eta$ values has been achieved in this case.

Considering the other discharge types, the most promising results on the $CO_2$ conversion have been obtained so far in an atmospheric DBD discharge using the effect of power modulation (open square in **Figure 10**), thus one more time underlining significance of this effect for better

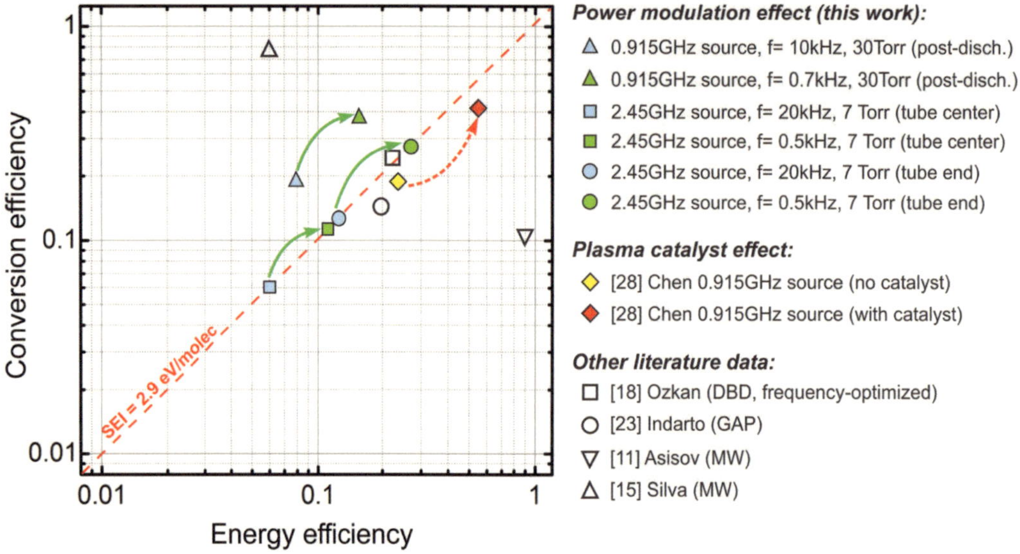


**Figure 10.** The optimized values of the $CO_2$ conversion efficiency ($\chi$) and energy efficiency ($\eta$) obtained in this work in both the 2.45-GHz and 0.915-GHz plasma sources and compared to the most competitive literature data. The gain obtained by the power modulation and the catalysis effects in our case is indicated by solid and dashed arrows (respectively).

plasma-assisted $CO_2$ conversion. Somewhat lower conversion efficiency has been achieved in an atmospheric GAP case, as studied by Indarto et al. [23] (open circle). On the other hand, two other examples related to the low-to-moderate pressure microwave discharges, one representing high conversion efficiency, but rather low energy efficiency attained as a result of applying high SEI [15] (open up-triangle), and the other representing the well-known work of Asisov et al. [11] where a supersonic gas flow enabled high energy efficiency (open down-triangle). These two examples are shown in order to illustrate the well-known $\chi$-$\eta$ tradeoff, especially evident in the nonoptimized discharge cases. In this regard, the results presented in this work clearly demonstrate the importance of the discharge tuning, as one of the possible ways to partially overcome this tradeoff, simultaneously achieving high conversion and energy efficiencies of $CO_2$ conversion.

## 5. The perspectives

Based on the obtained experimental results as well as on the comparison with the literature data, it is clear that there is a definite room for further improvement of the $CO_2$ conversion in low-temperature microwave plasma. It should be noted that the improvements achieved by tuning only the plasma parameters, without changing the energy expenses in the whole system, are implied in this case. The beneficial effect of modulation (i.e., timely interruption) of the electric power delivered to the discharge is already evident based on the results described in this chapter. At the same time, there are still few physical parameters that should be considered critical for the further maximization of the plasma-assisted $CO_2$ conversion. The main factors influencing these improvements are summarized in **Figure 11**, and can be subdivided into three following groups:

- Power-related parameters
- The parameters related to gas mixture and gas dynamics
- Plasma catalysis

Among the power-related parameters, the plasma pulse duty ratio, as well as the gas residence time (related to the gas velocity in the discharge as well as to the gas pumping speed in the whole system), requires deeper investigation. One may expect an essential improvement of the $CO_2$ conversion efficiency as a result of more careful optimization of the gas residence time in the active zone, along with the optimization of the plasma pulse duty ratio, in order to fully utilize the resonant effects related to the relevant energy transfer processes described in this work.

The second group includes the gas mixture control as well as the gas flux optimization including gas expansion effects. The influence of the gas mixture in the microwave plasma has already been studied recently showing a beneficial effect for $CO_2$ decomposition [14]. The optimization of the gas flux dynamics, e.g., via the gas expansion for the sake of lowering kinetic gas temperature, on the other hand, represents a powerful way to control the V-T

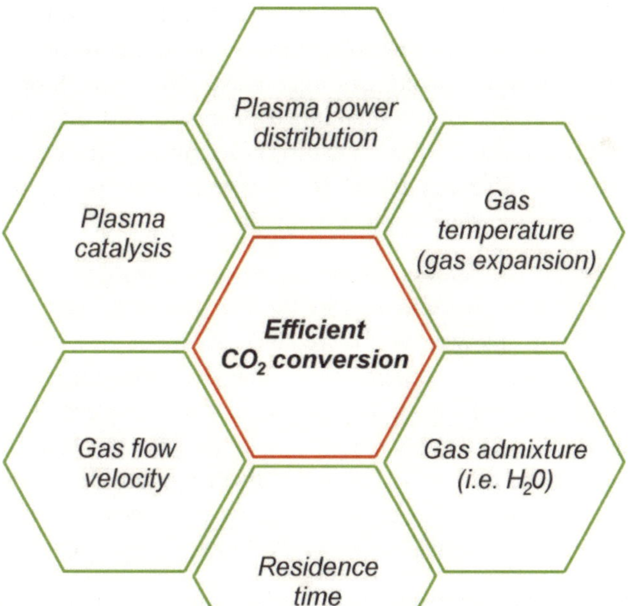


**Figure 11.** Summary of the critical factors influencing $CO_2$ conversion in microwave flowing gas discharges.

transfer rates, and thus the vibrational distributions of the molecules of interest. This work is supposed to be assisted by plasma modeling targeted at clarification of the numerous important kinetic processes.

Finally, the plasma catalysis activities are supposed to be especially beneficial since an essential gain in the $CO_2$ conversion efficiency can be achieved in this case. As shown recently [14, 28], the regenerative properties of the Ni-based catalysts are the microwave plasma combined with their long lifetime [51] enable roughly a twofold increase in the overall $CO_2$ conversion and energy efficiencies. The questions related to the utilization of plasma catalysis for improvements of the $CO_2$ conversion efficiency are discussed in the following chapter.

## 6. Conclusions

Several physical aspects related to the power modulation during the plasma-assisted conversion of $CO_2$ in the low-to-moderate pressure flowing gas microwave discharges are discussed. The beneficial effects of controlling the plasma pulse frequency (in the vicinity of 1 kHz or lower in our case) for increasing the $CO_2$ conversion efficiency are clearly demonstrated.

In particular, it was shown that by tuning the plasma pulse repetition frequency, the $CO_2$ conversion and energy efficiencies can be improved several times. In the 2.45-GHz plasma system, a fourfold overall improvement has been registered. Based on the estimations of the characteristic time corresponding to the relevant energy transfer processes for the studied pulse frequency range (0.5–30 kHz), it was concluded that the electron-vibrational as well as vibrational-translational energy transfer mechanisms are mainly responsible for the optimization of $CO_2$ conversion in a flowing gas microwave discharge. A resonant-like relation between the characteristic time of the mentioned processes and the plasma pulse-on time may lead to either high or low $CO_2$ decomposition efficiency, depending on the plasma pulse frequency. The maximum efficiency found so far corresponds to the frequencies of about 0.5–0.8 kHz depending on the studied plasma source.

At the same time, the gas temperature in the discharge active zone (i.e., near the excitation point) is supposed to be strongly influenced by the V-T and e-T energy transfer, as well as by the O atoms recombination at low repetition frequencies. Under these conditions at the beginning of the postdischarge, the gas temperature is somewhat reduced, as compared to the high frequency case. At high pulse frequency, the gas temperature in the active zone is getting lower by few hundreds degree K, presumably mainly due to the reduced O recombination (which is an exothermic process), whereas at the end of the discharge tube, it elevates following the O atom density, which is higher in this case. Numerous processes involving O atom kinetics should be studied by modeling, however, before clarifying this question fully.

In addition to the modulated microwave power delivery to the discharge, studied in this work, the other ways of optimization of the MW plasma-based $CO_2$ conversion are suggested. Among them are the optimizations related to the gas residence time, plasma pulse duty ratio, gas admixture, gas expansion in the discharge tube, and plasma catalysis. The results described in this chapter, combined with the suggested ways for the $CO_2$ conversion

http://dx.doi.org/10.5772/67875

improvements are supposed to be useful for the industry-oriented applications dealing with a local plasma-assisted $CO_2$ conversion.

## Acknowledgements

This study is supported by Belgian Government through the "Pôle d'Attraction Interuniversitaire" (PAI, P7/34, "Plasma-Surface Interaction," PSI). In addition, the partial support of the following projects should be acknowledged: the "REFORGAS GreenWin" project, grant No. 7267, (for N.B., G.C., T.G.), as well as the Portuguese FCT, under the projects UID/FIS/50010/2013 and PTDC/FIS-PLA/1420/2014 (for T.S.).

## Author details

Nikolay Britun[1]*, Guoxing Chen[1,2], Tiago Silva[3], Thomas Godfroid[4], Marie-Paule Delplancke-Ogletree[2] and Rony Snyders[1,4]

*Address all correspondence to: nikolay.britun@umons.ac.be

1 Chimie des Interactions Plasma Surface, University of Mons, Belgium

2 4MAT, Université Libre de Bruxelles, Belgium

3 Instituto de Plasmas e Fusão Nuclear, University of Lisbon, Portugal

4 Materia Nova Research Center, Belgium

## References

[1] R. M. Cuéllar-Franca and A. Azapagic, J. $CO_2$ Utilization **9**, 82 (2015).

[2] A. P. H. Goede, EPJ Web Conf. **98**, 7002 (2015).

[3] N. Britun, T. Minea, S. Konstantinidis, and R. Snyders, J. Phys. D Appl. Phys. **47**, 224001 (2014).

[4] A. Bogaerts, C. De Bie, R. Snoeckx, and T. Kozák, Plasma Process. Polym. DOI: 10.1002/ppap.201600070 (2016).

[5] A. A. Fridman and L. A. Kennedy, *Plasma Physics and Engineering* (Taylor and Francis, New York, 2011).

[6] A. A. Fridman, *Plasma Chemistry* (Cambridge University Press, New York, 2005).

[7] V. D. Rusanov, A. A. Fridman, and G. V Sholin, Sov. Phys. Uspekhi **24**, 447 (1981).

[8] V. N. Ochkin, *Spectroscopy of Low Temperature Plasma* (Wiley-VCH, Weinheim, 2009).

[9] Y. A. Lebedev, J. Phys. Conf. Ser. **257**, 12016 (2010).

[10] V. A. Legasov, *Nuclear – Hydrogen Energy and Technology. Vol. 1.* (Atom-Izdat, Moscow, 1978).

[11] R. I. Asisov, A. K. Vakar, V. K. Jivotov, et al., Proc. USSR Acad. Sci. **271**, 94 (1983).

[12] A. Fridman, S. Nester, L. A. Kennedy, A. Saveliev, and O. Mutaf-Yardimci, Prog. Energy Combust. Sci. **25**, 211 (1999).

[13] T. Godfroid, J. P. Dauchot, and M. Hecq, Surf. Coatings Technol. **200**, 649 (2005).

[14] G. Chen, N. Britun, T. Godfroid, V. Georgieva, R. Snyders, and M.-P. Delplancke, J. Phys. D Appl. Phys. **50**, 084001 (2017).

[15] T. Silva, N. Britun, T. Godfroid, and R. Snyders, Plasma Sources Sci. Technol. **23**, 025009 (2014).

[16] R. Aerts, W. Somers, and A. Bogaerts, Chem Sus Chem **8**, 702 (2015).

[17] S. Paulussen, B. Verheyde, X. Tu, C. De Bie, T. Martens, D. Petrovic, A. Bogaerts, and B. Sels, Plasma Sources Sci. Technol. **19**, 034015 (2010).

[18] A. Ozkan, T. Dufour, T. Silva, N. Britun, R. Snyders, F. Reniers, and A. Bogaerts, Plasma Sources Sci. Technol. **25**, 055005 (2016).

[19] A. Ozkan, T. Dufour, T. Silva, N. Britun, R. Snyders, A. Bogaerts, and F. Reniers, Plasma Sources Sci. Technol. **25**, 025013 (2016).

[20] D. Mei, Y.-L. He, S. Liu, J. Yan, and X. Tu, Plasma Process. Polym. **13**, 544 (2016).

[21] X. Duan, Y. Li, W. Ge, and B. Wang, Greenh. Gases Sci. Technol. **5**, 131 (2015).

[22] T. Nunnally, K. Gutsol, A. Rabinovich, A. Fridman, A. Gutsol, and A. Kemoun, J. Phys. D. Appl. Phys. **44**, 274009 (2011).

[23] A. Indarto, D. R. Yang, J.-W. Choi, H. Lee, and H. K. Song, J. Hazard. Mater. **146**, 309 (2007).

[24] S. R. Sun, H. X. Wang, D. H. Mei, X. Tu, and A. Bogaerts, J. $CO_2$ Utilization **17**, 220 (2017).

[25] L. F. Spencer and A. D. Gallimore, Plasma Chem. Plasma Process. **31**, 79 (2011).

[26] L. F. Spencer and A. D. Gallimore, Plasma Sources Sci. Technol. **22**, 015019 (2013).

[27] S. Mahammadunnisa, E. L. Reddy, D. Ray, C. Subrahmanyam, and J. C. Whitehead, Int. J. Greenh. Gas Control **16**, 361 (2013).

[28] G. Chen, T. Godfroid, N. Britun, V. Georgieva, M.-P. Delplancke, and R. Snyders, Appl. Catal. B. Environ. *Submitted* (2017).

[29] T. Kozák and A. Bogaerts, Plasma Sources Sci. Technol. **23**, 045004 (2014).

[30] J. A. Blauer and G. R. Nickerson, Rep. Ultrasystems Inc. (1973).

[31] K. Kutasi, V. Guerra, and P. Sá, J. Phys. D. Appl. Phys. **43**, 175201 (2010).

[32] J. G. Parker, Phys. Fluids **2**, 449 (1959).

[33] T. Silva, N. Britun, T. Godfroid, J. van der Mullen, and R. Snyders, J. Appl. Phys. **119**, 173302 (2016).

[34] SIGLO Database, www.lxcat.net, Retrieved Nov 24, 2016.

[35] R. C. Millikan and D. R. White, J. Chem. Phys. **39**, 3209 (1963).

[36] B. Chapman, *Glow Discharge Processes* (John Wiley & Sons, Ltd, New York, 1980).

[37] Phelps Database, www.lxcat.net, Retrieved Nov 24, 2016.

[38] M. Moisan and Z. Zakrzewski, J. Phys. D. Appl. Phys. **24**, 1025 (1991).

[39] G. Chen, V. Georgieva, T. Godfroid, R. Snyders, and M. P. Delplancke-Ogletree, Appl. Catal. B Environ. **190**, 115 (2016).

[40] T. Silva, N. Britun, T. Godfroid, and R. Snyders, Opt. Lett. **39**, 6146 (2014).

[41] G. F. Kirkbright and M. Sargent, *Atomic Absorption and Fluorescence Spectroscopy* (Academic Press, London, 1974).

[42] K. Niemi, V. S. Der Gathen, and H. F. Döbele, J. Phys. D. Appl. Phys. **34**, 2330 (2001).

[43] J. Rosell, J. Sjöholm, M. Richter, and M. Aldén, Appl. Spectrosc. **67**, 314 (2013).

[44] R. W. B. Pearse and A. G. Gaydon, *The Identification of Molecular Spectra*, 4th ed. (Chapman and Hall, London, 1976).

[45] C. Rond, A. Bultel, P. Boubert, and B. G. Chéron, Chem. Phys. **354**, 16 (2008).

[46] A. V Eremin and V. S. Ziborov, J. Appl. Mech. Tech. Phys. **34**, 752 (1993).

[47] N. Britun, M. Palmucci, S. Konstantinidis, M. Gaillard, and R. Snyders, J. Appl. Phys. **114**, 013301 (2013).

[48] G. Chen, T. Silva, V. Georgieva, T. Godfroid, N. Britun, R. Snyders, and M. P. Delplancke-Ogletree, Int. J. Hydrogen Energy **40**, 3789 (2015).

[49] T. Silva, N. Britun, T. Godfroid, and R. Snyders, Plasma Process. Polym. DOI: 10.1002/ppap.201600103 (2016).

[50] D. Marinov, D. Lopatik, O. Guaitella, M. Hübner, Y. Ionikh, J. Röpcke, and A. Rousseau, J. Phys. D. Appl. Phys. **45**, 175201 (2012).

[51] M. Zhang, D.-G. Cheng, and Y.-P. Zhang, Prepr. Pap.-Am. Chem. Soc., Div. Fuel Chem. **49**, 188 (2004).

Chapter 2

# Role of Plasma Catalysis in the Microwave Plasma-Assisted Conversion of $CO_2$

Guoxing Chen, Nikolay Britun, Thomas Godfroid,
Marie-Paule Delplancke-Ogletree and Rony Snyders

Additional information is available at the end of the chapter

http://dx.doi.org/10.5772/67874

### Abstract

Climate change and global warming caused by the increasing emissions of greenhouse gases (such as $CO_2$) recently attract attention of the scientific community. The combination of plasma and catalysis is of great interest for turning plasma chemistry in applications related to pollution and energy issues. In this chapter, our recent research efforts related to optimization of the conversion of $CO_2$ and $CO_2/H_2O$ mixtures in a pulsed surface-wave sustained microwave discharge are presented. The effects of different plasma operating conditions and catalyst preparation methods on the $CO_2$ conversion and its energy efficiency are discussed. It is demonstrated that, compared to the plasma-only case, the $CO_2$ conversion and energy efficiency can be enhanced by a factor of ~2.1 by selecting the appropriate conditions. The catalyst characterization shows that Ar plasma treatment results in a higher density of oxygen vacancies and a comparatively uniform distribution of NiO on the $TiO_2$ surface, which strongly influence $CO_2$ conversion and energy efficiencies of this process. The dissociative electron attachment of $CO_2$ at the catalyst surface enhanced by the oxygen vacancies and plasma electrons may explain the increase of conversion and energy efficiencies in this case. A mechanism of plasma-catalytic conversion of $CO_2$ at the catalyst surface in $CO_2$ and $CO_2/H_2O$ mixtures is proposed.

**Keywords:** green energy, $CO_2$ conversion, plasma catalysis, microwave discharge, $NiO/TiO_2$ catalyst

## 1. Introduction

One of the biggest concerns of the twenty-first century is the effect of global warming and its destructive consequences on the global ecosystem. Evidence indicated that the main cause of increased amounts of greenhouse gas emissions (such as $CO_2$) are the human activities related to the industrialization and burning of the fossil fuels. The process of conversion of $CO_2$ along with the other greenhouse gases into value-added chemicals or new fuels has attracted considerable attention because of the current background of fossil resources depletion and increase of $CO_2$ emissions. If a renewable (from sunlight or wind) or nuclear energy source drives the dissociation of $CO_2$ or other greenhouse gases, the renewable or nuclear energy can be stored as chemical energy in fuels. In this context, $CO_2$ reutilization to synthesize syngas, valuable fuels, or chemical compounds as well as pure $CO_2$ dissociation into CO and $O_2$ is of a special interest. Large efforts have been made to develop energy-efficient technologies [1–7].

Different techniques for conversion of $CO_2$ into value-added chemical compounds or fuels activated by heat or electricity have been developed and studied, namely thermolysis, photocatalysis, and electrochemical methods [1–5]. The $CO_2$ thermal splitting is thermodynamically and energetically favorable only at very high temperatures (1400–1800°C) and the conversion efficiency is very low. A prototype of $CO_2$ thermolysis chamber using only solar energy was developed by Traynor and Jensen [2]. The measured net conversion of $CO_2$ to CO was close to 6% under optimal conditions. The maximum conversion of solar to chemical energy was 5%. Recently, photocatalytic conversion of $CO_2$ using solar energy has attracted more and more attention. However, one of the greatest drawbacks is the low conversion efficiency at present. Moreover, its application is limited due to the numerous disadvantages [3, 4]: (1) inefficient reactor system, (2) catalyst activation needing long periods of irradiation, and (3) inefficient exploitation of visible light. High-temperature electrolysis is another promising method since it can effectively convert the electrical energy into chemical energy (78–87%) [5]. Electrolysis of $CO_2$ in the liquid phase has been investigated extensively while only few studies have been reported for electrolysis of $CO_2$ in the gas phase. Much effort should be made to overcome the drawbacks such as short-term stability of electrolytes, using expensive or scarce materials, and whole system heating.

Plasma technology represents an alternative approach to the above-mentioned methods. The nonthermal plasma technology is considered to be an attractive alternative to other (classical) technologies for converting inert carbon emissions (such as $CO_2$) into valuable fuels and chemicals, due to its nonequilibrium characteristics, environmental-safety, scalability, and low power requirements [6]. The fundamental properties and usefulness of the nonthermal plasmas for conversion of carbon dioxide as well as for related plasma technologies are extensively reviewed by Fridman [6], Zou and Liu [7], and by the other numerous authors. Nonthermal plasmas including dielectric barrier discharges (DBDs) [8–13], microwave discharges (MWs) [14–21], gliding arc plasmatrons (GAPs) [22–24], glow discharges [25], radio frequency (RF) discharges [26], and corona discharge [27] have been investigated for $CO_2$ conversion so far. However, a known trade-off between the energy efficiency and conversion efficiency is observed in plasma-assisted dissociation of pure $CO_2$. In order to overcome this trade-off, a plasma-assisted catalytic process with plasma activation of the catalysts to increase the conversion efficiency while maintaining high energy efficiency has been proposed. The combination of heterogeneous catalysis and plasma activation, known as plasma catalysis, has attracted increasing interest [28–31]. Plasma-catalytic conversion of $CO_2$ is a complex and challenging process involving a large number of physical and chemical reactions. The efficiency of this process can be controlled by means of the plasma parameters themselves, as well as by the catalyst properties. This suggests that more systematic studies on both the plasma effects and the chemical effects of the catalyst are necessary.

It is well known that the vibrational excitation is considered to be the most efficient elementary process in $CO_2$ plasmas to stimulate the endothermic $CO_2$ decomposition via the so-called vibrational ladder climbing process [6]. Due to its high degree of nonequilibrium, microwave plasma excites highly vibrational states of $CO_2$ molecules, which are energy-efficient for $CO_2$ decomposition. In this regard, combining catalysis with microwave plasmas could be especially promising, as the catalyst surface can reach a somewhat elevated temperature and possibly induce a synergy. Up to now, there are only a limited number of attempts for combining catalyst with microwave plasma [15, 17, 19–21].

The present Chapter critically summarizes the plasma-catalyst phenomena for optimization of the microwave plasma-assisted $CO_2$ conversion focusing on $NiO/TiO_2$ catalysts. It is focused on the investigation of several plasma-catalytic effects and their influences including the catalyst preparation methods, plasma activation and NiO content, aiming at enhancing our understanding of the plasma-catalytic $CO_2$ conversion process.

## 2. Brief theoretical background

The excitation of both CO{_2} and water vapor is required for effective dissociation. The high energy plasma electrons have the ability to efficiently excite and dissociate $CO_2$ and $H_2O$ via collisional processes. The dissociation of a $CO_2$ molecule is represented by the following global reaction [6]:

$$CO_2 \rightarrow CO + \tfrac{1}{2}O_2,\ \Delta H = 2.9\,\text{eV/molecule} \quad (1)$$

The process given by Eq.1 can be represented by an electron impact $CO_2$ dissociation [6]:

$$CO_2 \rightarrow CO + O,\ \Delta H = 5.5\,\text{eV/molecule} \quad (1a)$$

following either by O recombination into $O_2$ [Eq. (1b)] or by the reaction with a ground state (but vibrationally excited) $CO_2$ molecule [Eq. (1c)]:

$$M + O + O \rightarrow O_2 + M\,(\text{M is a particle}) \quad (1b)$$

$$O + CO_2 \rightarrow CO + O_2,\ \Delta H = 0.3\,\text{eV/molecule.} \quad (1c)$$

Besides this, CO can be easily converted into $CO_2$ with the production of $H_2$ in the presence of water, which is known as the water-gas shift reaction (WGSR):

$$CO + H_2O \rightleftarrows CO_2 + H_2,\ \Delta H = -0.4\,\text{eV/molecule (forward direction).} \quad (2)$$

Rather than disassociating the $CO_2$ directly, $H_2$ can reduce $CO_2$ to CO via the reverse direction depending on the reaction conditions. The analysis of the $CO_2$ and $H_2O$ dissociation thermodynamics shows that the WGSR is favored in the forward direction if the temperature is less than about 827°C. Above that value the WGSR is favored in the reverse direction [5].

The decomposition of a $H_2O$ molecule is given by the following reaction [6]:

$$H_2O \rightarrow H_2 + \frac{1}{2}O_2, \Delta H = 2.6\,\text{eV/molecule} \tag{3}$$

Traditionally, in order to characterize the process efficiency, two main parameters reflecting the energy efficiency and conversion efficiency are used: conversion efficiency and energy efficiency. By definition [4], the energy efficiency $\eta$ of dissociation process is given by

$$\eta = \chi \frac{\Delta H}{\text{SEI}} \tag{4}$$

where $\Delta H$ is the dissociation enthalpy of the global reaction [Eq. (1)], that is, 2.9 eV/molecule in the case of $CO_2$, $\chi$ is the $CO_2$ conversion efficiency [Eq. (5)], and SEI is the specific energy input per molecule (in eV).

The conversion efficiency of $CO_2$ is defined based on the following ratio:

$$\chi = \frac{\text{moles of } CO_2 \text{ converted}}{\text{moles of } CO_2 \text{ supplied}} \tag{5}$$

In our case, $\chi$ is calculated by comparing the gas chromatography (GC) $CO_2$ peak area without and with the presence of plasma discharge, as described in Ref. [15]. The specific energy input per molecule, SEI, is given by the ratio of the discharge power ($P$) to the gas flow rate ($F$) through the discharge volume and it can be expressed in eV per molecule.

$$\text{SEI} = \frac{P}{F} \tag{6}$$

## 3. Experimental part

### 3.1. The microwave discharge reactor

**Figure 1** shows a schematic diagram of the surface-wave microwave set-up used in this work. The discharge was sustained in a quartz tube (14 mm in diameter and 31 cm in length) by an electromagnetic wave with a filling frequency of 915 MHz. The quartz tube was additionally cooled by 5°C oil. In order to minimize the reflected microwave power, the impedance of the waveguide was automatically adjusted by a three-stub tuning system. The reflected power measured for each condition was always below the detection limit of the measuring probe and thus considered negligible. The $CO_2$ or the other gases (Ar, $O_2$, $H_2$, or $H_2O$) used in our study are injected from the top of the system. The gas flow rate was regulated by electronic mass flow controllers ranging between 0 and 6 standard liter per minute (slm). The water vapor was generated in an Omicron technologies vaporization system producing a vapor flow rate between 0 and 3 slm. The obtained discharge pressure was ranging from 30 to 60 Torr. A reactor containing the catalyst was placed at the end of the plasma quartz tube, about 2 cm below the end of the tube. The sine microwave waveform (at 915 MHz) is modulated in square pulses using a dedicated power supply. Accordingly, the kHz range frequency is related to the

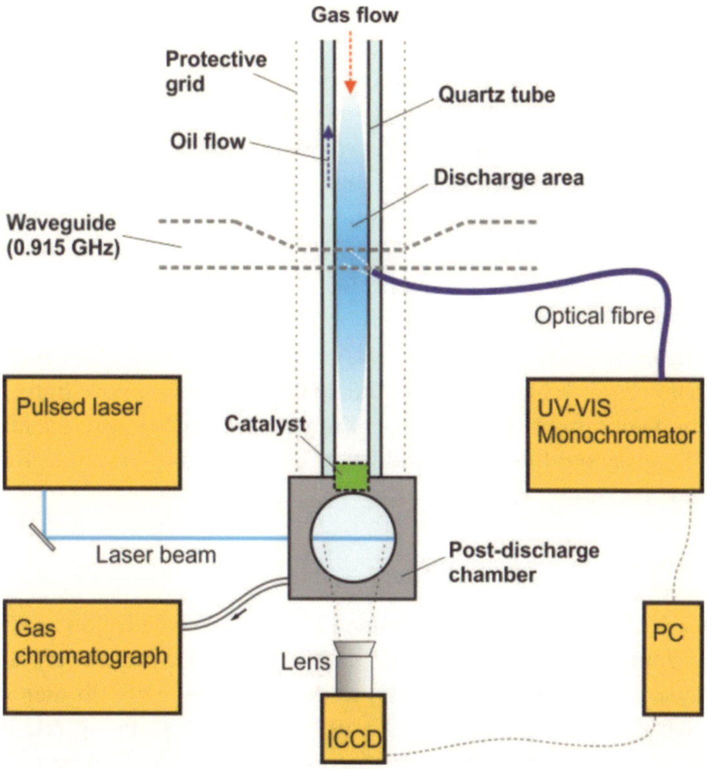


**Figure 1.** Schematic representation of surface-wave microwave set-up with the discharge diagnostic facilities used in this work.

modulation frequency, on top of the base frequency. In the present study, a 50% duty cycle was used: a pulse duration of 300 μs and a period of 600 μs (pulse frequency = 1.67 kHz). A more detailed description of the microwave set-up can be found elsewhere [15, 19, 32].

### 3.2. Characterization of plasma and dissociation products

The composition of the postdischarge was analyzed by a Bruker-450 gas chromatograph (GC) system equipped with a carbon molecular sieve column and a molecular sieve 5A column in series and connected to a thermal conductivity detector (TCD). Argon was used as a carrier gas in order to facilitate detection of $H_2$. $CO_2$, CO, $H_2$, and $O_2$ were successfully detected by GC. No hydrocarbons were detected by this current system. The microwave discharge was also characterized by optical emission spectroscopy (OES), which is a useful technique to study the large variety of chemical processes in plasma. The light emission from diatomic molecules is generally represented by a set of ro-vibrational emission bands in the studied spectral range (~300–800 nm) [33]. The vibrational populations, gas temperature, densities of the various particles and radicals, vibrational temperature, and molecular dissociation degree can be obtained through the analysis of these emission bands (as well as the atomic emission lines). The corresponding experimental data is given in the previous Chapter and also can be found elsewhere [17, 18]. The present Chapter is mainly dedicated to the effects of plasma catalysis.

### 3.3. Catalyst preparation

In the present work, we consider two different types of the catalyst treatment: conventional calcination and plasma treatment. Pellets (6 mm diameter and 1 mm thick) made from pure anatase powder (Sigma-Aldrich) were used as a catalyst support. We use plasma calcination as a replacement of the conventional calcination in the plasma-treatment process. Three types of catalyst were prepared by plasma processing and are denoted in this work as: NiO/$TiO_2$ ($O_2$), NiO/$TiO_2$ (Ar), and NiO/$TiO_2$ ($CO_2$) which corresponds to the catalysts pretreated with pure $O_2$ plasma, pure Ar plasma, and pure $CO_2$ plasma, respectively. The catalysts prepared by conventional calcination are referred to as, NiO/$TiO_2$ (Air-C), NiO/$TiO_2$ (Ar-C), for calcination in air and argon, respectively. A catalyst (NiO/$TiO_2$ (Ar-P)) was also studied with an Ar plasma for a longer time (similar to conventional calcination). The summary of the preparation conditions and Brunauer-Emmett-Teller (BET) surface areas is given in **Table 1**. More details about these two catalysts and the preparation can be found in the corresponding literature studies [15, 21].

### 3.4. Catalyst characterization techniques

The surface area of the catalyst was determined by nitrogen adsorption at a temperature of 77 K using BET analysis. At the same time, the analysis of the catalyst crystalline structure was conducted using X-ray diffraction (XRD) on the catalyst powder (Bruker advanced X-ray diffractometer (40 kV, 40 mA) with a Cu K$\alpha$ (0.154 nm) at a scanning rate of 2°/min from 20 to 90°). TEM (transmission electron microscopy) (Philips CM 200) images were used to characterize the morphology and crystalline size. The Raman spectra were collected using a Bruker spectrometer equipped with a 532-nm argon ion laser as the excitation source (the laser power was 2 mW). The spectrum range was from 40 to 1000 $cm^{-1}$ and the spectral resolution was 0.5 $cm^{-1}$. Raman spectrometry was also used to detect the presence of defects in the crystal lattice. Ultra violet -to- visible (UV-VIS) diffuse reflectance spectra were obtained by a Cary spectrophotometer using $BaSO_4$ as a background at a resolution of 1 $cm^{-1}$. It was used

| **Catalysts** | **Description** | **BET surface area ($m^2/g$)** |
|---|---|---|
| $TiO_2$ | Catalyst support (pure anatase) | 12 |
| $TiO_2$ (Ar) | Ar plasma treated for 40 minutes | 12 |
| NiO/$TiO_2$ ($O_2$) | $O_2$ plasma treated for 40 minutes | 15 |
| NiO/$TiO_2$ ($CO_2$) | $CO_2$ plasma treated for 40 minutes | 19 |
| NiO/$TiO_2$ (Ar) | Ar plasma treated for 40 minutes | 19 |
| NiO/$TiO_2$ (Air-C) | Conventional calcination in air for 2 hours | 18 |
| NiO/$TiO_2$ (Ar-C) | Conventional calcination in Ar for 2 hours | 17 |
| NiO/$TiO_2$ (Ar-P) | Ar plasma treated for 2 hours | 19 |

**Table 1.** Summary of various catalysts and pure $TiO_2$ applied for plasma-assisted catalytic conversion of $CO_2$.

for measurements of the bandgap energy and the absorbance of ultraviolet light as a function of the wavelength.

## 4. Plasma-assisted catalytic conversion of $CO_2$

### 4.1. The influence of catalyst preparation method

In the field of catalysis, the nature of the catalyst in terms of its physical structure and chemical properties is a key factor for its effectiveness in a particular process. The activity of the adsorption sites provided by the catalyst determines the mechanisms that provide the lower energy reactive pathways [28]. The formation of these adsorption sites could be sensitively affected by catalyst preparation conditions, and the understanding of the chemical basis for this is essential. The nonthermal plasma, having highly energetic electrons and chemically reactive species (e.g., free radicals, excited atoms, ions, and molecules), may modify surface properties of the catalyst, enhance the dispersion of the supported metals, and even create disorder in their crystallite structure, depending on the nature of the gas phase, and the experimental treatments conditions. In this work, in order to study the relationship between the catalyst preparation method (its structure, morphology, and defect) and its efficiency in the $CO_2$ conversion, two different catalyst preparation methods, namely the conventional calcination and plasma treatment, were considered. Comparison of $CO_2$ conversion and energy efficiencies in the microwave discharge with and without catalysts as well as the effect of different preparation methods are investigated and presented in **Figure 2**. Thermally calcinated

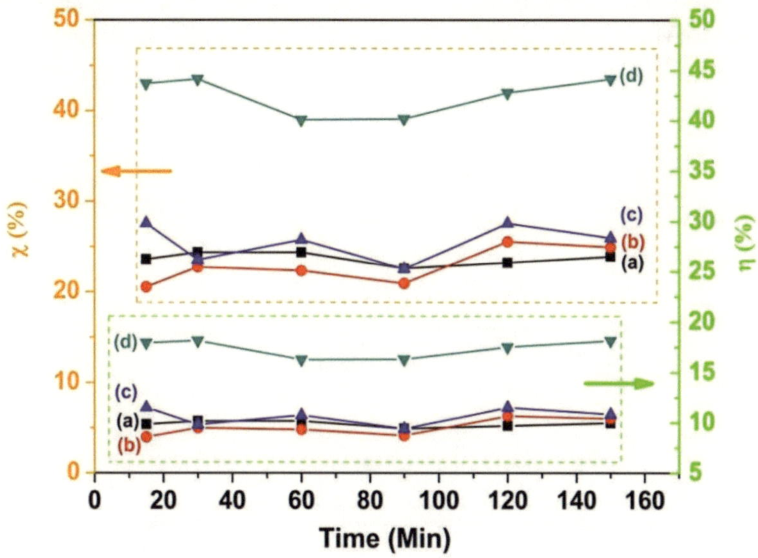


**Figure 2.** $CO_2$ conversion and energy efficiencies are shown for the NiO/$TiO_2$ catalysts prepared by different methods (conventional calcination vs Ar plasma treatment). (a) Plasma only, (b) NiO/$TiO_2$ (Air-C), (c) NiO/$TiO_2$(Ar-C), (d) NiO/$TiO_2$ (Ar-P). Flow rate = 2 slm; pressure = 30 Torr; SEI = 6.95 eV/molecule. pulse frequency = 1.67 kHz.

catalysts (NiO/$TiO_2$ (Air-C) and NiO/$TiO_2$ (Ar-C)) affect the $CO_2$ conversion rather insignificantly (in comparison with the plasma-only-assisted $CO_2$ dissociation). On the other hand, it is clear that the presence of plasma-activated catalyst (NiO/$TiO_2$(Ar-P) significantly enhances the $CO_2$ conversion and energy efficiencies compare to the plasma-only-assisted decomposition. The $CO_2$ conversion and energy efficiencies increase from 23 to 43% and 10 to 18%, respectively, compared to the plasma only case. For better understanding of the influence of the different preparation methods on chemical and physical properties of the catalysts, the chemical composition and morphology of the catalysts were studied by XRD, BET analysis, TEM, and Raman spectroscopy.

The catalyst characterization based on XRD and TEM results shows that an Ar plasma treatment results in a very uniform distribution of nickel oxide on the $TiO_2$ and the possibility of a higher $Ni^{2+}$ concentration into the $TiO_2$ lattice. Raman spectroscopy was used in order to access the phase constitution and lattice defects of the catalyst support ($TiO_2$). All observed Raman peaks correspond to the characteristic Raman modes of the anatase $TiO_2$ structure [34, 35] (**Figure 3**). The peak shift in Raman spectra indicates that the vibrations of the $TiO_2$ lattice are considerably affected by the presence of Ni ions. The largest shift and broadening occur for the Ar plasma-treated catalyst (NiO/$TiO_2$ (Ar-P)), as shown in **Figure 3**. The $E_g$ peaks are associated with the symmetric stretching vibration mode of O-Ti-O in $TiO_2$. This mode is very sensitive to local oxygen coordination surrounding the metal ions. According to a previous detailed analysis of Raman spectra of $TiO_2$ and doped-$TiO_2$ samples [15], the broadening and shifting of the main bands can be attributed to the presence of oxygen vacancies. These results indicate that plasma treatment can induce higher density of oxygen vacancies compared to the conventional calcination. To get a valuable insight into the catalyst activation process, a two-dimensional model of Ar plasma was developed in the modeling framework PLASIMO [36–38]. The results from the simulation imply the effective catalyst activation by the Ar metastable species (4s $1s_5$ and 4s $1s_3$ states) at the catalyst chamber region [21]. The complex environment encountered

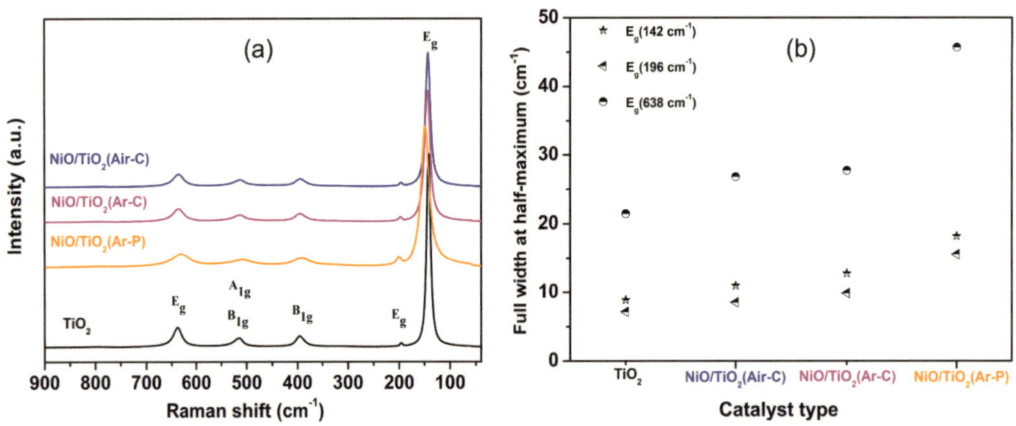


**Figure 3.** (a) Raman spectra of conventional calcined and plasma-treated catalysts and pure $TiO_2$; (b) full-width at half-maximum (FWHM) in $E_g$ modes in the different catalysts and $TiO_2$ support.

in plasma could lead to the unusual catalyst structure, formation of smaller particles, better dispersion, and stronger interaction between the catalyst and the support.

### 4.2. The influence of plasma gas for catalyst activation

As mentioned above, plasma plays an important role in tailoring the properties of a catalyst by the selective doping of a material with heteroatoms and its subsequent modification of the bandgap of a material which can affect both activity and selectivity [28]. However, modifications of the catalyst depend on the nature of plasma gas and experimental treatment conditions. The influence of preactivation of the catalyst using three different gases, namely $O_2$, Ar, or $CO_2$, on the $CO_2$ conversion and energy efficiencies as well as on the chemical and physical properties is studied in this section. The $CO_2$ conversion and energy efficiencies are not altered significantly in the presence of $TiO_2$ (Ar) support (without doping agent) compared to the plasma-only $CO_2$ decomposition case, as shown in **Figure 4**. The presence of NiO/$TiO_2$ (Ar) significantly increases the $CO_2$ conversion efficiency and energy efficiency almost twice from 23 to 42% and 9.6 to 17.5%, respectively, comparing with the plasma-only experiment results. However, the $CO_2$ conversion and energy efficiencies were not significantly affected when NiO/$TiO_2$ ($O_2$) and NiO/$TiO_2$ ($CO_2$) catalysts are used. To understand this effect, the analysis of different catalysts and supports was undertaken by XRD, BET, Raman spectroscopy, and UV-VIS spectroscopy. Similar crystallite sizes (around 46 nm) were observed for pure $TiO_2$ support, $TiO_2$ (Ar) and NiO/$TiO_2$ catalysts prepared by plasma [15]. On the other hand, the lattice constant c increases slightly with the addition of NiO and the largest increase

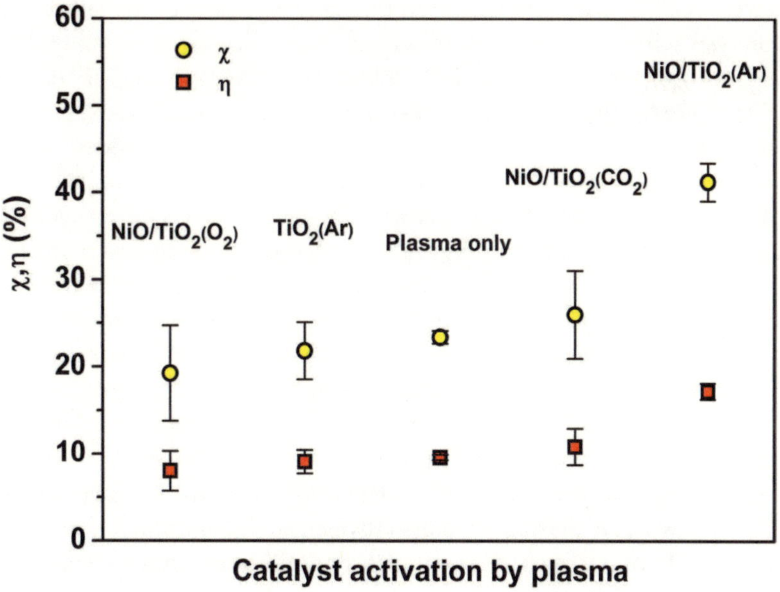


**Figure 4.** $CO_2$ conversion and energy efficiencies are shown for the NiO/$TiO_2$ catalysts prepared by plasma treatment with different gases ($O_2$, Ar, $CO_2$). Flow rate = 2 slm; pressure = 30 Torr; SEI = 6.95 eV/molecule. Pulse frequency = 1.67 kHz. Reprinted with permission from Ref. [15]. Copyright 2016 Elsevier.

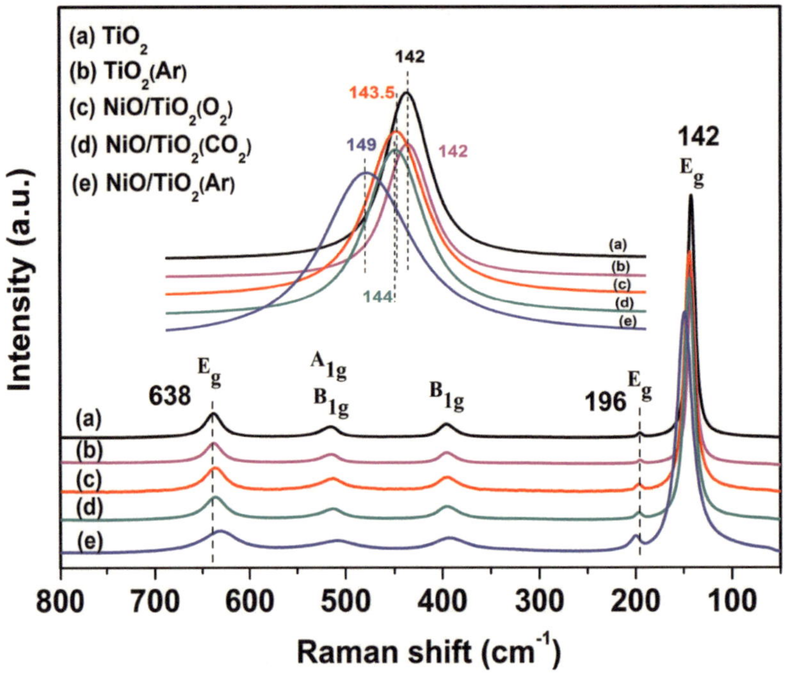


**Figure 5.** Raman spectra of $TiO_2$-supported NiO catalysts, $TiO_2$ (Ar), and pure $TiO_2$ support. The enlarged view of the intense $E_g$ peak is shown in the inset. Reprinted with permission from Ref. [15]. Copyright 2016 Elsevier.

is observed for NiO/$TiO_2$ (Ar) catalyst based on the comparison of the lattice values for the NiO/$TiO_2$ catalysts with pure $TiO_2$. The observed $TiO_2$ lattice expansion might imply that a few $Ni^{2+}$ ions are incorporated into the $TiO_2$ lattice ( the ionic radius of $Ni^{2+}$ (0.83 Å) is larger than that of $Ti^{4+}$ (0.75Å) [39]). As we have discussed in Section 4.1, the oxygen vacancies are probably mainly responsible for the blue and red shifts as well as for widening of the Raman peaks corresponding to $E_g$ modes. Together with the XRD results, we conclude the substitution of $Ti^{4+}$ ions by $Ni^{2+}$ distorts the $TiO_2$ lattice and generates oxygen vacancies to maintain charge neutrality. As it is noted from **Figure 5** that the largest shift and widening occur for the Ar plasma-treated NiO/$TiO_2$ catalyst. This indicates that the Ar plasma treatment leads to a higher concentration of oxygen vacancies in the presence of NiO. In order to study the doping effect on the catalyst optical properties, the bandgap energy $E_{bg}$ was calculated based on the UV-VIS spectra [15]. The obtained bandgap energies are 2.85, 2.79, 2.60, 2.49, and 2.17 eV for pure $TiO_2$, $TiO_2$ (Ar), NiO/$TiO_2$ ($O_2$), NiO/$TiO_2$ ($CO_2$), and NiO/$TiO_2$ (Ar), respectively. These results are consistent with the previous findings that the increasing of oxygen vacancies results in a narrowing bandgap energy for $TiO_2$ [40–43]. The pure $TiO_2$ support and the $TiO_2$ support pretreated with Ar plasma were also investigated. No influence of Ar plasma pretreatment on the $TiO_2$ support surface area and its crystalline structure as well as on the defect formation has been found. Therefore, it proves that an important condition for the formation of oxygen vacancies by means of Ar plasma pretreatment is the presence of $Ni^{2+}$ which leads to incorporation of the $Ni^{2+}$ ions in the $TiO_2$ lattice.

http://dx.doi.org/10.5772/67874

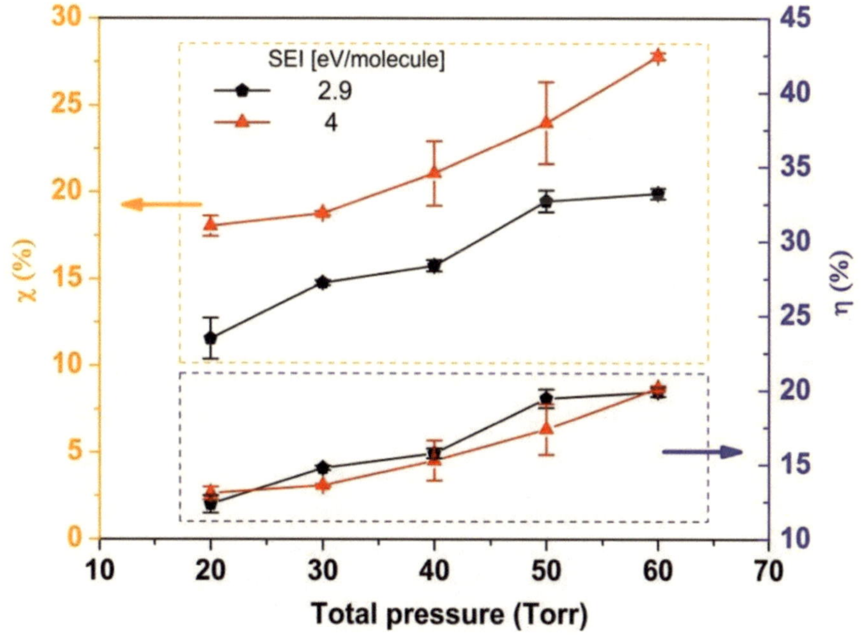


**Figure 6.** Evolution of the conversion efficiency and energy efficiency of $CO_2$ as a function of the discharge pressure at different SEI.

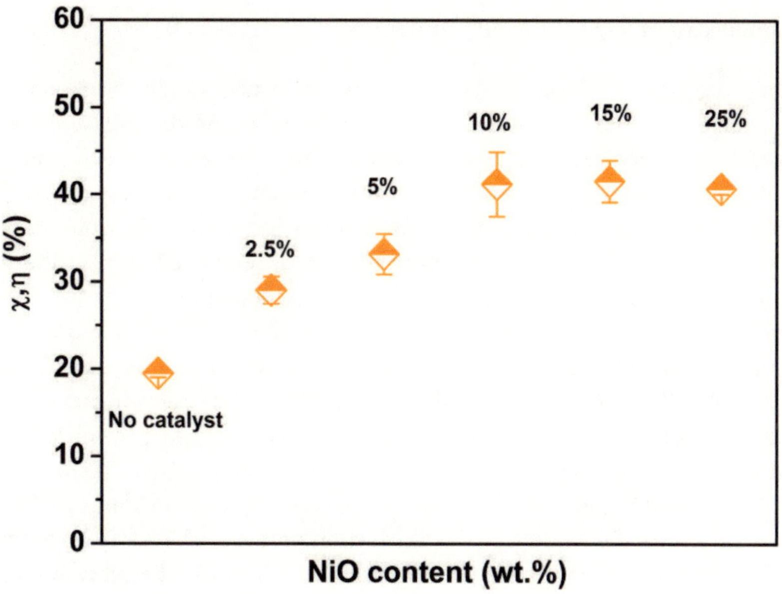


**Figure 7.** Dependence of the $CO_2$ conversion and energy efficiencies on the NiO content in the NiO/$TiO_2$ catalysts treated by Ar plasma. Flow rate = 5 slm; pressure = 60 Torr; SEI = 2.9 eV/molecule. Pulse frequency = 1.67 kHz.

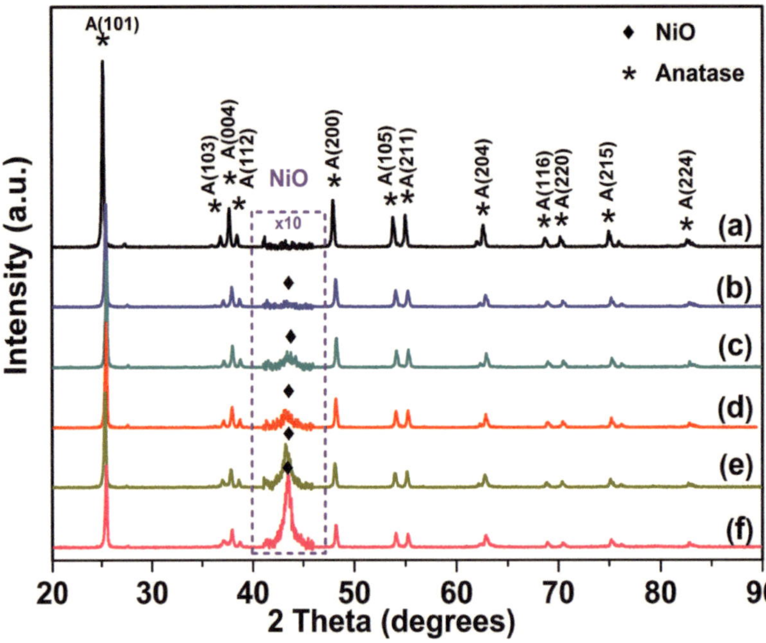


**Figure 8.** XRD patterns of plasma-treated catalysts with different NiO loading and pure $TiO_2$: (a) $TiO_2$, (b) 2.5 wt% NiO, (c) 5 wt% NiO, (d) 10 wt% NiO, (e) 15 wt% NiO, (f) 25 wt% NiO. The peaks, associated with the anatase phase and denoted with * in the XRD pattern of the pure $TiO_2$ (anatase) support, are observed with the similar intensity in all samples.

### 4.3. The influence of pressure and NiO content

First, the total pressure effect on the $CO_2$ decomposition was studied in the absence of catalyst in the range of 20–60 Torr at different SEI (**Figure 6**). As the pressure increases, our experimental data clearly show an improvement in $CO_2$ conversion, which is in good agreement with published results obtained in microwave plasma at low pressure [6]. These pressure-dependent results can be understood based on the increase of electron-neutral collision frequency in the plasma phase which favors the vibrational excitation of the asymmetric vibrational mode of $CO_2$ and leads to higher $CO_2$ conversion efficiency [6]. In addition, $\eta$ drops when SEI increases whereas at the same time $\chi$ increases. This trade-off between $\chi$ and $\eta$ has been observed in other $CO_2$ experiments as well [6, 10]. From this result it is clear that $\chi$ does not increase to the same extent as SEI does. The effect of nickel contents on the $CO_2$ conversion and energy efficiencies was investigated at 60 Torr. Increasing the nickel content from 2.5 to 10 wt% results in an increase of the $CO_2$ conversion and energy efficiencies up to 42% (**Figure 7**). However, at higher nickel content, there is no significant change in the conversion of $CO_2$. **Figure 8** shows the X-ray powder diffraction patterns of $NiO/TiO_2$ catalysts for different NiO contents. For nickel contents ranging from 2.5 to 10 wt%, very weak diffraction peaks were observed at $2\theta = 43.5^{\circ}$, corresponding to the (200) crystal planes of face-centered cubic NiO lattice. As a result of further increase of the NiO content, NiO diffraction peaks become clearly visible. This increase in the NiO content does not increase the population of

$Ti^{4+}$ substituted by $Ni^{2+}$ in the $TiO_2$ lattice. Further investigations are necessary to obtain a better understanding of the role of NiO.

## 5. Discussion

Among several factors giving significant improvements in both $\chi$ and $\eta$ values, the following two groups are of special interest: the nature of plasma activation gas and NiO content on the $TiO_2$ surface. Compared to the plasma-only case, the $CO_2$ conversion and energy efficiencies were multiplied by a factor of 2.1 by selecting the appropriate catalyst preparation methods (plasma) and the NiO content. In agreement with previous study [19] on the $CO_2$ conversion efficiency and the corresponding relative Raman shifts of the three $E_g$ modes observed in NiO/$TiO_2$ ($O_2$, Ar, $CO_2$), NiO/$TiO_2$ (Air-C), NiO/$TiO_2$ (Ar-C), NiO/$TiO_2$ (Ar-P), and in $TiO_2$ (Ar) (shown in **Figures 3** and **5**), the present results indicates that the higher activity observed on Ar plasma-treated NiO/$TiO_2$ catalysts could be linked to the formation of surface oxygen vacancies.

It is well known that photocatalysts can be activated through the formation of electron-hole ($e^-$-$h^+$) by photoillumination of the catalyst surface. $TiO_2$-based catalysts are extensively studied for $CO_2$ photoreduction because of the adequate bandgap of $TiO_2$ (3.2 eV for anatase phase). Liu et al. [44] have shown that the mechanism of activation and subsequent reduction of $CO_2$ involves the participation of protons and electron transfer. However, with a pure $TiO_2$ (Ar) support, no difference in the $CO_2$ conversion efficiency compared to plasma only case was found, as shown in **Figure 8**. Hence, the observed improvements in our work could not be explained by the possible photoreduction of $CO_2$ induced by the plasma radiation.

Nowadays, the fabrication of oxygen-deficient surfaces has increasing interest, which is an important strategy for improving the photocatalytic $CO_2$ dissociation [44–46]. Oxygen vacancy (Vo) has been considered as the active site for the adsorption and activation of $CO_2$. The electron attachment to a $CO_2$ molecule adsorbed by an oxygen vacancy leads to the formation of a transient negative ion $CO_2^-$ that decomposes to give CO and fill the Vo with an oxygen atom between a $Ni^{2+}$ and $Ti^{4+}$ ions (the oxygen atom is $O_{br}$) [15]. The threshold energy for this one-electron process is found to be 1.4 eV [44, 47]. The dissociative electron attachement (DEA) of $CO_2$ observed on the catalytic surface can occur also in the gas phase [6]. However, the maximum value of the DEA cross-section is low ($\sim10^{-22}$ $m^2$) and the threshold energy of 5–10 eV [6] is high compared to the threshold energy of 1.4 eV [47] for the DEA at the catalyst surface. So, the DEA process for $CO_2$ occurs more easily for adsorbed $CO_2$ on the $TiO_2$ defective surface than for $CO_2$ in the gas phase. As the activity of our catalyst remains constant over time, a regeneration of Vo is necessary [15].

A study of $CO_2$ conversion efficiency in different gas mixtures in the presence of NiO/$TiO_2$ (Ar-P) has also been investigated [19, 21]. It was found that the $CO_2$ conversion efficiency was also enhanced independently of the mixture when NiO/$TiO_2$ (Ar-P) is used. A synergy between the catalyst and plasma is clearly demonstrated. The $CO_2$ conversion and energy efficiencies in 90% $CO_2$–10% $H_2O$ increase from 30 to 52% and 12.5 to 22%, respectively,

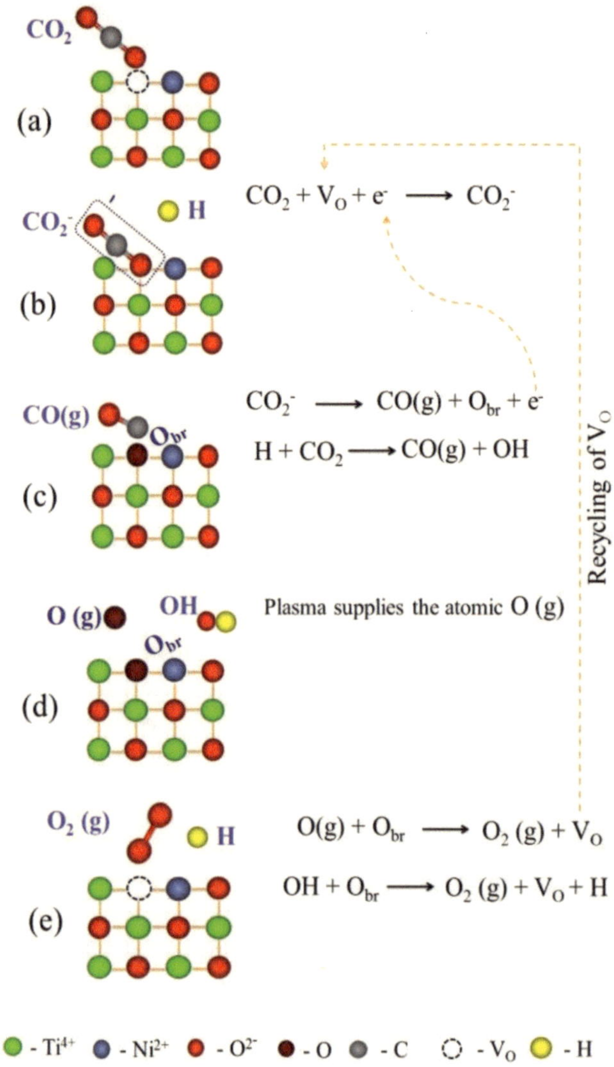


**Figure 9.** A tentative mechanism of plasma-catalytic $CO_2$ decomposition on the catalyst surface.

when the activated catalyst is combined with the plasma. We found previously that oxygen vacancies enhanced the $CO_2$ conversion in the plasma-catalytic dissociation of pure $CO_2$ [15]. The electrons supplied by the plasma enhance the dissociative attachment of $CO_2$ on the oxygen-deficient catalyst surface. We suggest that the following processes can be distinguished on the catalyst surface in the plasma-catalytic $CO_2/H_2O$ conversion (**Figure 9**). $CO_2$ could be reduced to CO via reaction with H radical or direct dissociation by healing the oxygen vacancy sites [1, 48, 49]. The active species OH and adsorbed oxygen atom react with each other to form atomic hydrogen and $O_2$, resulting in the regeneration of oxygen vacancy. The oxygen vacancy can also be regenerated via the recombination on the surface of a bridging oxygen atom with a gaseous oxygen atom [15, 50, 51]. The presence of atomic

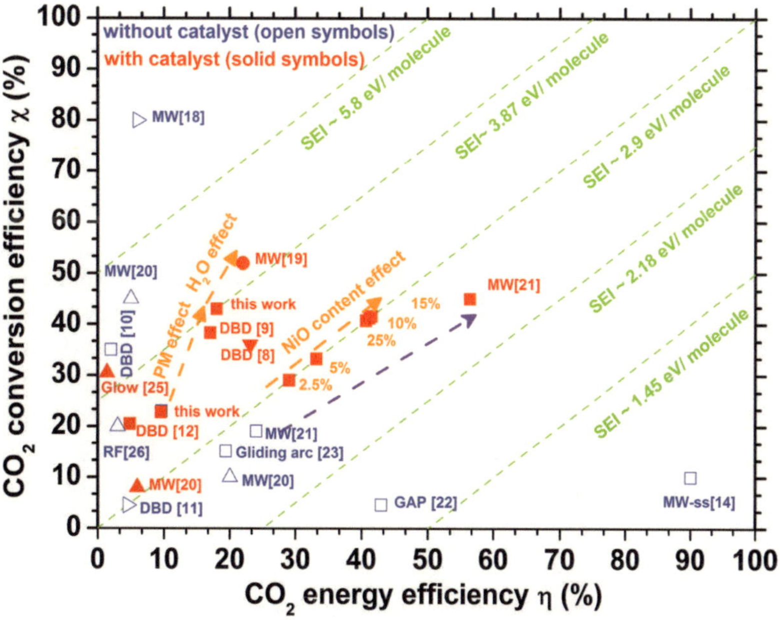


**Figure 10.** Comparison of the conversion and energy efficiencies for CO2 with different types of non-thermal plasmas: □, ■ - Pure CO2; Δ, ▲ - CO2-Ar mixture; ● - CO2-H2O mixture; ▼ CO2-H2O -Ar mixture; ▷-CO2-N2 mixture. PM stands for Preparation Method, MW-ss stands for a Microwave discharge with supersonic gas flow. The SEI values are given for reference.

O peaks and water dissociation product (OH and H) peaks in the emission spectrum of $CO_2$/$H_2O$ microwave discharge was detected and reported in our previous study [17]. The plasma supplies energy to the catalyst surface and thus additionally enhances the recombination process. The regeneration of oxygen vacancies can be confirmed by our experimental results [19, 21]. The $CO_2$ conversion and energy efficiencies remain more or less constant over time, at least for 150 minutes of continuous operation. Hence, a synergistic effect between the plasma and the NiO/$TiO_2$(Ar-P) catalyst leads to significantly improved $CO_2$ conversion and energy efficiencies, which can be attributed to the dominant surface reactions driven by the plasma.

A two-dimensional $\chi$-$\eta$ chart compares the energy efficiency for $CO_2$ conversion using different types of nonthermal plasmas (**Figure 10**). The influence of the NiO content, as well as of the other effects studied (such as $H_2O$ effect) in this and previous work [19, 21] is clearly visible. Selected results on $CO_2$ conversion taken from the literature are given for comparison. The data points corresponding to our study look very promising. The highest energy efficiency of about 90% was reported by Asisov et al. in microwave plasma in supersonic flow [14]. The corresponding $CO_2$ conversion was only about 10%, which is significantly lower than that (42%) obtained in this work for pure $CO_2$ decomposition with a catalyst. The highest reported conversion efficiency was about 80% while the energy efficiency was less than 6% [18]. A balance between $CO_2$ conversion and energy efficiencies in the plasma processing of $CO_2$ is especially important for the development of an efficient and cost-effective process. The combination of the microwave plasma and the Ar plasma-treated NiO/$TiO_2$ catalyst leads to significant enhancement in the $CO_2$ conversion and energy efficiencies of the plasma process, as well as the balance between them.

## 6. Conclusions

This work critically summarizes the effects of various plasma operating conditions and catalyst preparation methods on the $CO_2$ conversion and its energy efficiency in a surface-wave sustained microwave discharge. As a result, one can conclude that the catalyst preparation method has a significant impact on the chemical and physical properties of the catalysts, which in turn strongly influence $CO_2$ conversion and energy efficiencies. Ar plasma treatment is supposed to result in a higher density of oxygen vacancies and a very favorable distribution of nickel oxide on the $TiO_2$ surface comparing with the classical calcination in air or in argon.

The oxygen vacancies are the key factor explaining the catalytic activities in $CO_2$ decomposition. These vacancies are stabilized by the presence of $Ni^{2+}$ ions in the anatase lattice. The plasma pretreatment allows to both create oxygen vacancies, and to incorporate the nickel into the $TiO_2$ lattice to stabilize these vacancies. The dissociative electron attachment of $CO_2$ at the catalyst surface enhanced by the oxygen vacancies and plasma electrons can explain the observed increase of $CO_2$ conversion as well as the energy efficiency. A mechanism is proposed that explains the observed plasma-catalyst synergy, which has led to improved $CO_2$ conversion and energy efficiencies.

At the same time, there is still a room for further improvement of the $CO_2$ conversion and energy efficiencies through the optimization of the plasma parameters (e.g., high pressure and high flow rate) and the modification of catalysts (e.g., loading different metal nanoparticles on $TiO_2$). In addition, it is very promising to explore the possibility of $H_2O$-$CO_2$ as an alternative system to conversion of carbon dioxide to value-added compounds, such as syngas, methanol, and formic acid. We expect that the results presented in this study will provide useful insights into the plasma-assisted $CO_2$ conversion in the presence or absence of catalysts, which may be used for greenhouse gas conversion in industry.

## Acknowledgements

The authors acknowledge the financial support from the network on Physical Chemistry of Plasma Surface Interactions-Interuniversity Attraction Poles phase VII project (http://psi-iap7.ulb.ac.be/), supported by the Belgian Federal Office for Science Policy (BELSPO). G. Chen is grateful to Violeta Georgieva, Tiriana Segato, Patrizio Madau, and Löic Malet (Université Libre de Bruxelles) and to Sylvain Deprez (Materia Nova Research Center) for their valuable help in catalyst characterization and discussion. N. Britun and T. Godfroid acknowledge the support of the "REFORGAS GreenWin" project (grant No. 7267).

## Author details

Guoxing Chen[1,2*], Nikolay Britun[2], Thomas Godfroid[3], Marie-Paule Delplancke-Ogletree[1] and Rony Snyders[2,3]

*Address all correspondence to: guoxchen@ulb.ac.be

1 4MAT, Free University of Brussels, Belgium

2 Chemistry of Plasma Surface Interactions, University of Mons, Belgium

3 "Materia Nova" Research Center, Belgium

## References

[1] Rayne, S. Carbon Dioxide Splitting: A Summary of the Peer-Reviewed Scientific Literature. Available from Nature Precedings, 2008. DOI:10.1038/npre.2008.1741.1.

[2] Traynor A J, Jensen R J. Ind. Eng. Chem. Res. 2002; **41:** 1935–1939.

[3] Hu B, Guild C, Suib S L. J. $CO_2$ Util. 2013; **1**: 18–27.

[4] Stechel E B, Miller J E. J. $CO_2$ Util. 2013; **1**: 28–36.

[5] Graves C, Ebbesen S D, Mogensen M, Lackner K S. Renew. Sust. Energ. Rev. 2011; **15**: 1–23.

[6] Fridman A. Plasma Chemistry. Cambridge University Press, New York, 2008.

[7] Zou J, Liu C. Utilization of Carbon Dioxide Through Nonthermal Plasma Approaches. Wiley-VCH Press, ISBN: 978-3-527-32475-0, 2010. DOI: 10.1002/9783527629916.ch10.

[8] Mahammadunnisa S, Reddy L, Ray D, Subrahmanyam C, Whitehead J C. Int. J. Greenh. Gas Con. 2013; **16**: 361–363.

[9] Mei D, Zhu X, Wu C, Ashford B, Williams P P, Tu X. Appl. Catal., B. 2016; **182**: 525–532.

[10] Aerts R, Somers W, Bogaerts A. Chem Sus Chem. 2015; **8**: 702–716.

[11] Snoeckx R, Heijkers S, Wesenbeeck K Van, Lenaerts S, Bogaerts A, Energy Environ Sci., 2016; **9**: 999–1011.

[12] Yu Q, Kong M, Liu T, Fei J, Zheng X. Plasma Chem Plasma Process. 2012; **32**: 153.

[13] Ozkan A, Dufour T, Bogaerts A, Reniers F. Plasma Sources Sci. Technol. 2016; **25(4)**: 045016.

[14] Asisov R I, Givotov V K, Krasheninnikov E G, Potapkin B V, Rusanov V D, Fridman A. Sov. Phys., Doklady 1983; **271**: 94.

[15] Chen G, Georgieva V, Godfroid T, Snyders R, Delplancke-Ogletree M P. Appl. Catal. B: Environ. 2016; **190**: 115–124.

[16] Van Rooij G J, van den Bekerom D C M, Den Harder N, Minea T, Berden G, Bongers W A, Engeln R, Graswinckel M F, Zoethout E, Van de Sanden M C M. Faraday Discuss. 2015; **183**: 233–248.

[17] Chen G, Silva T, Godfroid T, Godfroid T, Britun N, Snyders R, Delplancke-Ogletree M P, Int. J. Hydrogen Energy. 2016; **40**: 3789–3796.

[18] Silva T, Britun N, Godfroid T, Snyders R. Plasma Sources Sci Technol. 2014; **23**: 025009.

[19] Chen G, Britun N, Godfroid T, Godfroid T, Snyders R, Delplancke-Ogletree M P. J. Phys. D: Appl. Phys. 2017; **50**: 084001.

[20] Spencer L F, Gallimore A D. Plasma Sources Sci. Techenol. 2013; **22**: 015019.

[21] Chen G, Godfroid T, Britun N, Georgieva V, Delplancke-Ogletree M P, Snyders R. Appl. Catal. B: Environ. 2016. *Submitted.*

[22] Nunnally T, Gutsol K, Rabinovich A, Fridman A, Gutsol G, Kemoun A. J. Phys. D: Appl. Phys. 2011; **44**: 274009

[23] Indarto A, Yang D R, Choi J W, Lee H, Song H K, J. Hazard. Mater. 2007; **146**: 309–315.

[24] Sun S R, Wang H X, Mei D H, Tu X, Bogaerts A, J. $CO_2$ Util. 2017; **17**: 220–234.

[25] Brock S L, Marquez M, Suib S L, Hayashi Y, Matsumoto H. J. Catal. 1998; **180**: 225–233.

[26] Spencer L F, Gallimore A D. Plasma Chem. Plasma Process. 2011; **31**: 79–89.

[27] Mikoviny T, Kocan M, Matejcik S, Mason N J, Skalny J D. J. Phys. D: Appl. Phys. 2004; **37**: 64.

[28] Chung W C, Chang M B. Renew. Sustain. Energy Rev. 2016; **62**: 13–31.

[29] Whitehead J C. J. Phys. D: Appl. Phys. 2016; **49**: 243001.

[30] Chen H L, Lee H M, Chen S H, Chao Y, Chang M B. Appl. Catal., B, 2008; **85**: 1–9.

[31] Neyts E C, Ostrikov K, Sunkara M K, Bogaerts A. Chem. Rev. 2015; **115**: 13408.

[32] Godfroid T, Dauchot J P, Hecq M, Surf. Coatings Technol. 2005; **200**: 649.

[33] Herzberg G. Molecular Spectra and Molecular Structure: Spectra of Diatomic Molecules. New York: Litton Educational Publishing, 1950.

[34] Ohsaka T, Izumi F, Fujiki Y. J. Raman Spectrosc. 1978; **7**: 321.

[35] Choudhury B, Choudhhury A. Mater. Chem. Phys. 2012; **132**: 1112–1118.

[36] Georgieva V, Berthelot A, Silva T, Kolev S, Graef W, Britun N, Chen G, Van der Mullen J, Godfroid T, Mihailova D, Dijk J V, Snyders R, Bogaerts A, Delplancke-Ogletree M P. Plasma Processes Polym. 2016, DOI:10.1002/ppap.201600185

[37] The Plasimo code website: http://plasimo.phys.tue.nl/

[38] Van Dijk J, Peerenboom K S C, Jimenez-Diaz M, Mihailova DB, Van der Mullen J J A M, J. Phys. D: Appl. Phys. 2009; **42**: 194012.

[39] Shannon R D. Acta Cryst A. 1976; **32**: 751–767.

[40] Suib S. New and Future Developments in Catalysis. Elsevier, Amsterdam, 2013.

[41] Wang Y, Zhang R, Li J,Li L, Lin S. Nanoscale Res. Lett. 2014; **9**: 46.

[42] Bharti B, Kumar S,Lee H N, Kumar R. Sci. Rep. 2016; **6**: 32355.

[43] Kamarulzaman N, Kasim M F, Rusdi R. Nanoscale Res Lett. 2015; **10**: 346.

[44] Liu L, Li Y. Aerosol. Aerosol. Air. Qual. Res. 2014; **14**: 453–469.

[45] Rasko J, Solymosi F. J. Phys. Chem. 1994; **98**: 7147–7152.

[46] Liu L, Zhao C, Li Y. J. Phys. Chem. C. 2012; **116**: 7904–7912.

[47] Lee J, Sorescu D C, Deng X, J. Am. Chem. Soc. 2011; **133**: 10066–10069.

[48] Tahir M, Tahir B, Amin N A S, Muhammad A. Energy Convers. Manag. 2016; **119**: 368–378.

[49] Liu L, Zhao H, Andino J M, Li Y. ACS Catalysis 2012; **2**: 1817–1828.

[50] Crane P J, Dickens P G, Thomas R E. Trans. Faraday Soc. 1967; **63**: 693–700.

[51] Kovalev V L, Krupnov A A, Pogosbekyan M Y, Sukhanov L P, Acta Astronautica. 2011; **68**: 686–690.

Section 2

# Biomass Green Process

Chapter 3

# Catalytic Conversions of Biomass-Derived Furaldehydes Toward Biofuels

Shun Nishimura and Kohki Ebitani

Additional information is available at the end of the chapter

http://dx.doi.org/10.5772/67805

**Abstract**

Upgrading of biomass resources toward high-energy compounds (biofuel) is a crucial technology for sustainable development because utilizations of biomass resources can contribute to the low $CO_2$ emission on the basis of carbon neutral concept. In this chapter, recent advances on catalytic hydrogenation and hydrogenolysis of biomass-derived furaldehydes, dehydration products of saccharides, for example, called as hydroxymethylfuran (HMF) and furfural, toward biofuels over heterogeneous catalytic system are introduced. Some approaches on mechanistic study and reactor design are also mentioned in this chapter.

**Keywords:** biomass, furaldehydes, biofuel, hydrogenation, hydrogenolysis

## 1. Introduction

In the past decades, owing to the rise of a living standard through social infrastructure development in the world, the energy demand growth has been stronger with increase of global gross domestic product (GDP), and it would rise further 30% till 2040 [1, 2]. While concerns on global warming derived from $CO_2$ emission have been debated in Intergovernmental Panel on Climate Change (IPCC) and Conference of the Parties (COP), they established global guidelines to participate in a sustainable development toward "zero emission." New energy sources and technologies have been current; however, it is still a great challenge to fabricate a low carbon society where high quality of life is constructed with low $CO_2$ emission. The rate of renewable energy including solar, geothermal, waste, wind, biofuels, wood, and hydroelectric power was only below 10% in the total primary energy consumption diagram still now [1].

A facile transformation of biomass-based resources is one of the crucial technologies to reduce the emission of $CO_2$ according to the carbon neutral concept; that is, the amount of $CO_2$ emission owing to biomass consumption (burning) is equal to the amount of $CO_2$ storage during its growth (photosynthesis). Thus, the balance of $CO_2$ concentration in the atmosphere on the basis of biomass utilization process is scarcely influenced. A lot of efforts have put into the investigations of new biomass transformation processes and derived technologies (e.g., see [3–8]). In this chapter, recent advanced catalytic transformations of biomass-based furaldehydes (furfurals) toward biofuels *via* hydrogenation/hydrogenolysis are introduced.

## 2. Transfer hydrogenation of furaldehydes to biofuels

A schematic reaction pathway for transfer hydrogenation/hydrogenolysis of biomass-based furaldehydes toward biofuels and chemicals is summarized in **Scheme 1**. Investigations on the synthesis of tetrahydrofuran (THF), tetrahydrofurfuryl alcohol (THFA) [9, 10], and 2-methyltetrahydrofuran (2-MTHF) [11] are attractive to serve biomass-based green solvent and precursor for aliphatic alcohols such as 1-butanol, 1,5-poentandiol, and 2-pentanol, respectively [12]. Moreover, furaldehyde-based diols of 2,5-bis(hydroxymethyl)furan (BHMF) [13–15] and 2,5-bis(hydroxymethyl)tetrahydrofuran (BHMTHF) [16–18] are the candidate for biomass-derived monomer source. Note, BHMF is also called as 2,5-dihydroxymethylfuran (DHMF) as a popular style. Accordingly, the strategy to rule such competitive reactions among reduction,

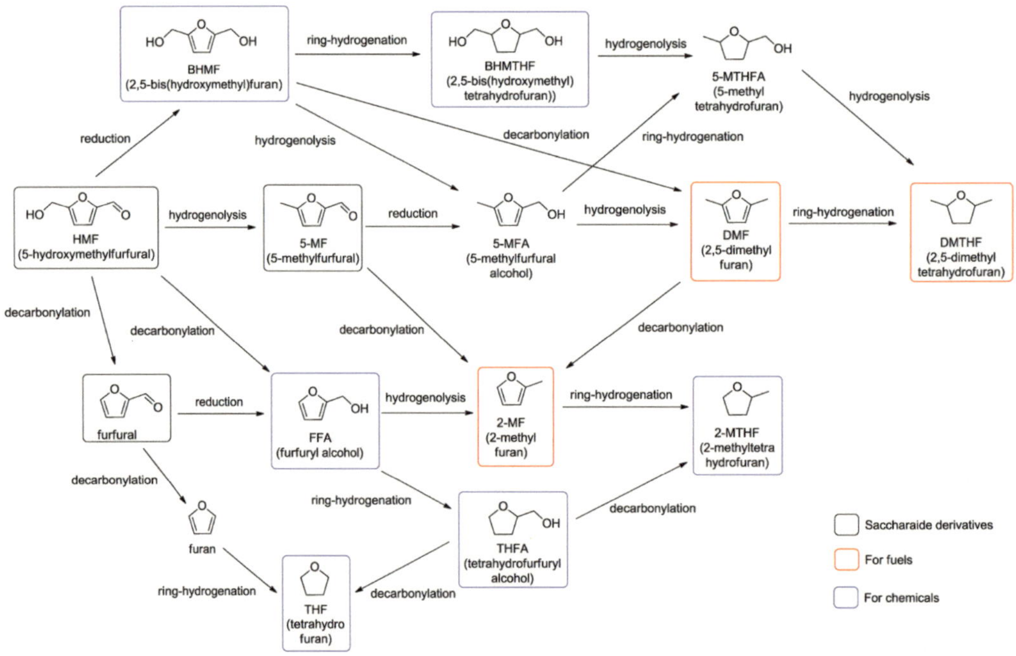


**Scheme 1.** Schematic reaction pathway for transfer hydrogenation/hydrogenolysis of biomass-based furaldehydes toward biofuels and chemicals.

ring-hydrogenation, hydrogenolysis, and decarbonylation is the key factor in the biorefinery on the transfer hydrogenation of furaldehydes.

The products of 2-methylfuran (2-MF), 2,5-dimethylfuran (DMF), and 2,5-dimethyltetrahydrofuran (DMTHF) have been counted on as biomass-based fuels (biofuels) as well as conventional biomass-derived transportation fuels such as bioethanol and biodiesel. These furan-based biofuels possess high-energy density [lower heating value (LHV): 28–32 MJ $L^{-1}$], low volatility [higher bp. (336–367 K) compared to ethanol], and good combustibility [research octane number (RON): 82–131]. Additionally, these are immiscible with water and stable compound in stage and thus would be an easier blender in gasoline than bioethanol. The DMF applications as automotive fuel have been challenged on a single-cylinder or multicylinder gasoline engines [19–21].

### 2.1. Synthesis of DMF by HMF hydrogenolysis

Synthesis of DMF from HMF is one of major researches for biofuel productions. First of all, studies on the synthesis of DMF from HMF hydrogenolysis (**Scheme 2**) have been introduced in the following five contents.

#### *2.1.1. At pressurized hydrogen condition*

Chidambaran and Bell provided 15% yield of DMF from hydroxymethylfuran (HMF) over Pd/C catalyst in ionic liquid/acetonitrile mixed solution under pressured $H_2$ (6.2 MPa) [22]. Recently, Saha and co-workers achieved 85% yield for DMF from HMF *via* BHMF intermediate with combined use of Pd/C and Lewis acid $ZnCl_2$ at lower pressurized $H_2$ (0.8 MPa) [23]. The improvement induced by $ZnCl_2$ agent was due to the facilitation on the rate-determinating step of BHMF to DMF *via* cleavage of C—O bonds. The synergism with $Zn^{2+}$ was varied by the metal center: Pd/C/$ZnCl_2$ was faster than Ru/C/$ZnCl_2$, and Ni/C shows poor synergism. Interestingly, BHMF was identified as an intermediate product during the reaction, and it was 52% yield at maximum with the Ru/C/$ZnCl_2$ system.

Ru-supported catalyst is examined by several groups in HMF hydrogenation at pressurized $H_2$ in recent days. Zu *et al.* investigated Ru/$Co_3O_4$ catalyst and conducted 93% yield for DMF at 403 K and 0.7 MPa $H_2$ [24]. In this system, Ru is responsible for hydrogenation, while $CoO_x$ works for the adsorption of hydrogenation product and then breaks the C—O bond. Nagpure *et al.* suggested Ru-NaY zeolite composed with Ru particles (Av. 2.8 nm) on the external surface served 78% yield of DMF under $H_2$ (1.5 MPa) at 493 K in a short duration of the reaction (1 h) [25]. According to the TOF value in DMF production at the

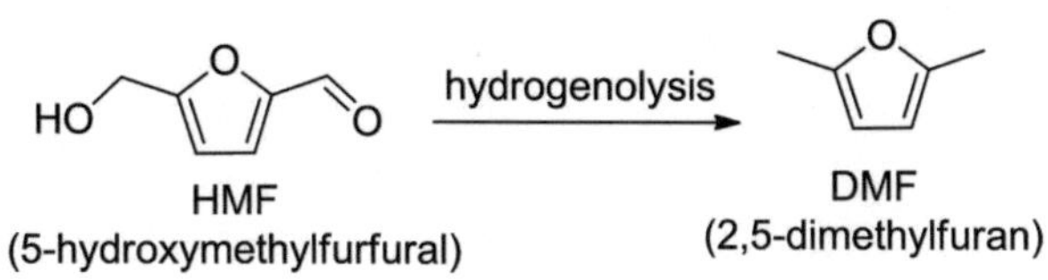


**Scheme 2.** Synthesis of DMF from HMF hydrogenolysis.

same condition, the reactivity of metal was expected as following order Ni (5.2 $h^{-1}$) < Cu (6.4 $h^{-1}$) < Au (16.4 $h^{-1}$) < Pt (60.8 $h^{-1}$) < Rh (80.2 $h^{-1}$) < Pd (98.6 $h^{-1}$) < Ru (156.0 $h^{-1}$), though the detailed information of size or surface area of other metals was not shown. The Lewis acidity of NaY zeolite was expected to increase deoxygenation ability and helped to improve the DMF yield in the reaction. On the other hand, Hu *et al.* gave 95% for DMF production over Ru/C catalyst at 473 K and 2 MPa $H_2$ [26]. The acid sites on carbon support supposedly promoted the hydrogenation of HMF to DMF. In addition, the authors also examined a combined use of ionic liquid (1-butyl-3-methylimidazolium chloride) and cellulose-derived sulfonated carbonaceous catalyst for HMF production from biomass-derived carbohydrates and successive hydrogenation of the extracted HMF to DMF over Ru/C catalyst, and then, more than 82% DMF yield was successfully obtained. It was generally known in the case of Ru catalyst that the reactivation by heating under $H_2$ flow is required to remove the depositions of high-molecular-weight by-products blocking the active Ru sites [26, 27].

In the case of Ni catalyst, the acidity of support is also expected to be the key function for deoxygenation step. Huang *et al.* explained the Ni particles mainly played a role in the hydrogenation step but had limited deoxygenation, whereas $W_2C$ particles mainly promoted the deoxygenation step of hydroxymethyl group but had limited hydrogenation ability. They have served 96% yield of DMF by using the synergy between Ni and $W_2C$ particles at 4 MPa $H_2$ [28]. It should be noted that the observed major intermediate was not BHMF but 5-MF in this study.

### 2.1.2. *With hydrogen donor agent*

Because the concentration of other undesirable byproducts *via* ring-open and ring hydrogenation such as BHMTHF, 5-methyltetrahydrofuran (5-MTHFA), DMTHF, and hexanediol increased at higher $H_2$ pressure [24, 29, 30], several approaches instead of the utilization of pressurized $H_2$ agent had been examined to build up further possible way in the catalytic system. Jae *et al.* applied 2-propanol as a hydride donor for Ru/C-catalyzed HMF hydrogenation and gave 81% yield for DMF at 463 K under pressurized $N_2$ (2.04 MPa) [27]. Two ethers *via* etherification of 2-propanol and BHMF or 5-MFA (denoted as ethers 2 and 3 in **Scheme 3**) were detected in the reaction, and the expected reaction network in the presence of 2-propanol

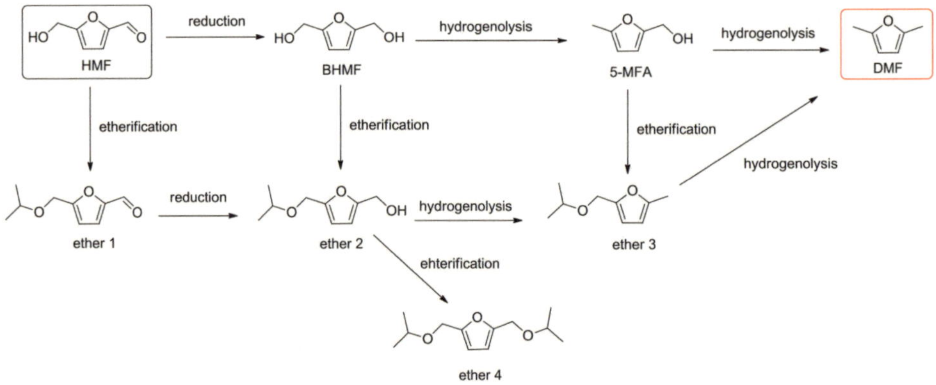


**Scheme 3.** Reaction network of the hydrogenation of HMF into DMF in the presence of 2-propanol.

http://dx.doi.org/10.5772/67805

is described in **Scheme 3**. Nagpure *et al.* demonstrated Ru-Mg(Al)O catalyst with 58% yield of DMF at 493 K for 4 h at 1 MPa $H_2$ assisted by hydrogen transfer from 2-propanol solvent [30]. Continuous-flow transfer hydrogenation/hydrogenolysis of HMF in 2-propanol yielded 72% DMF with Pd/$Fe_2O_3$ catalyst at 2.5 MPa pressure and 453 K through BHMF production at an initial stage (70% sel. at 50% conv.) [31]. Utilization of hydrogen donor agent would be a versatile strategy for the reaction.

One of the dramatic effects was found when the formic acid (HCOOH) agent was applied for the reaction. Rauchfuss *et al.* had surprisingly announced that decarbonylation of HMF to DMF (>95% yield) *via* BHMF formation route was proceeded by Pd/C catalyst and formic acid (10 equiv.) in refluxing THF/$H_2SO_4$ (0.13 equiv.) or dioxane (at 393 K) through diformate ester form (**Scheme 4**). The formic acid agent plays the multiroles, inhibits the decarbonylation and ring-hydrogenation, and serves mild hydrogen source and a precursor to formate ester [32, 33]. De *et al.* described Ru/C-catalyzed DMF formation from fructose with similar catalytic system, and 30% yield was obtained [34]. They also announced the positive effect of irradiation treatment (at 300 W) during reaction to significantly decrease the time for reaction.

### 2.1.3. *Under supercritical conditions*

As a different approach, Chatterjee *et al.* investigated the Pd/C catalyst under supercritical carbon dioxide ($scCO_2$) condition with various operation conditions [29]. Interestingly, they found that the $scCO_2$ pressure strongly contributed the selectivity for the HMF hydrogenation at 353 K and 1 MPa $H_2$ for 2 h (**Figure 1**). The selectivity for DMF was increased from 42 to 100% along with $scCO_2$ pressure increased from 4 to 10 MPa. 5-MTHFA was formed with a comparatively higher selectivity of 58% at 4 MPa and then decreased with pressure. At higher pressure (>14 MPa), DMTHF became the dominant product with about 70% selectivity. Controlled experiments with different $CO_2/H_2O$ molar ratio indicated that (*i*) in the absence of $H_2O$ or $CO_2$, the reaction proceeds through the path of HMF → 5-MFA → 5-MTHFA, and (*ii*) combined effect of $CO_2$ and $H_2O$ forced the reaction to move in the direct of HMF → DMF. Weak acidic condition derived from $CO_2$ dissolved in water would be one of the crucial factors

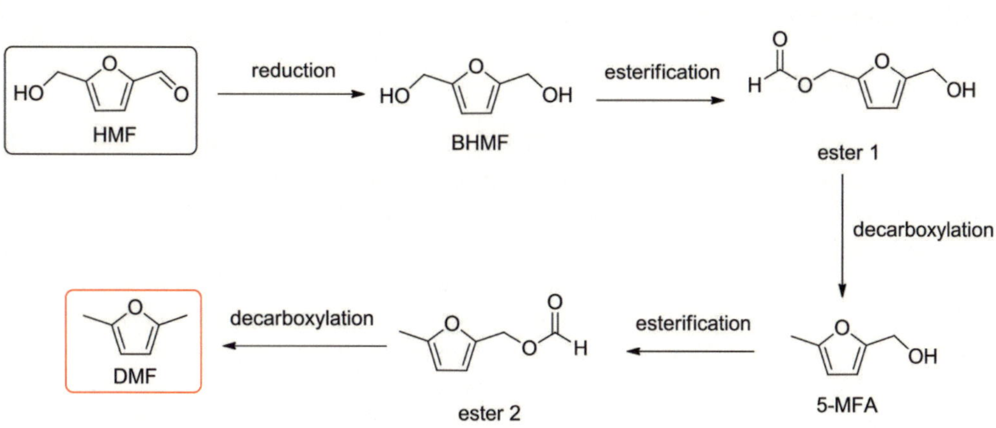


**Scheme 4.** Reaction network of the hydrogenation of HMF into DMF in the presence of formic acid.

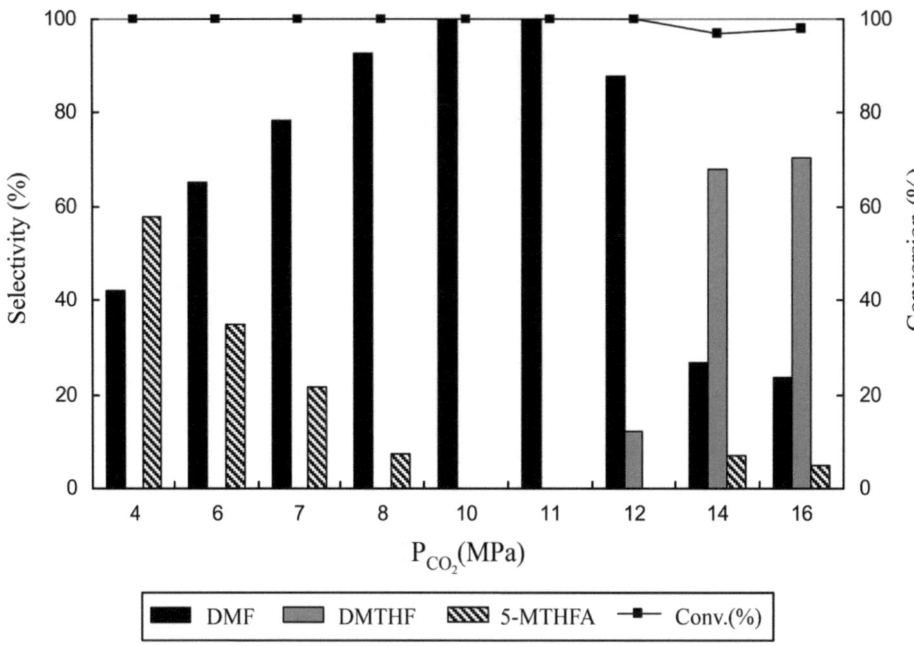


**Figure 1.** Effect of $CO_2$ pressure on the conversion and product profile. Reaction conditions: catalyst: substrate = 1:5, 353 K, 2 h, 1 MPa ($H_2$), water (1 mL). Reproduced with permission from The Royal Society of Chemical (RSC) of Ref. [29].

behind the hydrogenation of HMF to DMF. This catalytic system could be applied for furfural hydrogenation and 100% yield of 5-MF or 2-MTHF was achieved under optimized condition.

Hansen *et al.* demonstrated Cu-doped porous Mg-Al-$O_x$ (Cu-PMO) catalyst with supercritical methanol in a stainless steel bomb reactor [35]. According to their concept, the rapid deoxygenation of HMF in the presence of Cu-PMO drastically diminished undesired side reactions such as polymerizations and condensations. Total yield of three main products of DMF, DMTHF, and 2-hexanol was reached to be 61%. The maximum yield for DMF was 48% at 533 K for 3 h reaction.

### *2.1.4. Under ambient operation conditions*

In order to decrease the operation risks owing to the utilization of pressurized and/or high-temperature conditions, application of ambient condition for the target reaction is an ideal system. Moreover, as an additional issue, the undesired formation of insoluble humin is often observed and decreased the yield in furaldehyde utilizations at elevated reaction temperature [36]. One of impressive approaches under ambient operation conditions was reported by Bekkum *et al.* They have studied HMF hydrogenolysis under atmospheric $H_2$ at 333 K with Pd/C catalyst and gave 35.7% yield of DMF in 1-propanol solvent through propyl ether intermediates formation; the 2-methyl-5-(propoxymethyl)furan was detected in the initial stage with high yield (>80%) [37], while in 1,4-dioxane, mainly BHMF is formed with 80% yield at the same reaction condition.

Bimetallic catalysts for upgrading of biomass resources into high-value fuels and chemicals were the one of research interests in biomass conversion in the last decade [38–40]. Such movements motivated researchers to investigate bimetallic catalytic system for HMF hydrogenation with an atmospheric pressurized $H_2$.

Ebitani *et al.* prepared the PdAu bimetallic nanoparticles supported carbon ($Pd_xAu_y/C$) catalysts with different Pd/Au molar ratio (*x*/*y*) and applied to the HMF hydrogenation in an atmospheric $H_2$ in the presence of HCl agent [41]. In the monometallic catalysts, $Pd_{100}/C$ and $Au_{100}/C$ gave 60 and 0% yield for DMF yield, respectively. While, in the bimetallic PdAu series, increasing the Au contents in $Pd_xAu_y$ leads to increase the DMF yield till Pd/Au = 50/50 (presumption) and then slightly decrease; that is, positive effect on the coexistence of Au in Pd active center in the reaction was observed. With consideration of XRD experiments, they proposed that the highest yield for DMF, 96% at ambient $H_2$ condition with HCl agent at 333 K, would be served over the $Pd_{73}Au_{27}$ sites formed on as-prepared $Pd_{50}Au_{50}/C$ catalyst. It is expected that electronic and/or geometric changes in Pd, that is, the internal electronic transfer from Pd to another atom in agreement with the Pauling electronegativity scale and/or the modification of Pd-Pd atomic distance due to the insertion of another atom would play a crucial role for Pd-alloy mediated reactions [42, 43]; however, the detailed mechanism is a subject for a further study in HMF hydrogenation over PdAu/C catalyst.

### 2.1.5. *Over bimetallic active sites*

Approaches with bimetallic catalyst have been also investigated by other researches. Dumesic and coworkers studied the bimetallic Ru catalytic system for the reaction. The 46% yield of DMF from HMF was produced over RuSn/C catalyst at 473 K in the presence of lactone using a Parr reactor [44]. They also reported a systematic production of DMF using bimetallic RuCu/C catalyst in 1-butanol phase to afford 71% yield for 10 h at 393 K and 0.68 MPa $H_2$. This vapor-phase hydrogenolysis is capable for a two-step reaction from fructose connected with biphasic reactor for *in-situ* formed HMF feeder [45].

Interestingly, Yu *et al.* found the combination of low-selective Ni species in high conversion and inactive Fe species for HMF hydrogenation by alloying contributed to the significant catalytic performance at 3.0 MPa $H_2$ in *n*-butanol [46]. The selectivity to DMF showed a volcano-type behavior with increasing Fe loading, and the $Ni_2Fe_1/CNT$ catalyst gave the highest 91% yield for DMF at 473 K for 3 h. The author also reported 96% yield toward BHMF production at lower temperature (393 K); that is, the reaction temperature markedly affected the product distribution. It was expected that the longer C—O bond on Ni-Fe than that on Ni might facilitate the conversion of HMF to DMF as the main route. Wang *et al.* also induced the favorable adsorption of the C=O or C—O bond over that of the C=C by applying PtCo bimetallic sites. They achieved 98% yield of DMF formation from HMF at 453 K and 1 MPa $H_2$ over PtCo encapsulated in hollow carbon sphere (PtCo@HCS) catalyst in 1-butanol [47]. The author also indicated that the lower reaction temperature at 393 K served 70% yield of BHMF with PtCo@HCS catalyst.

A robust and highly active CuZn nanopowder (60 mesh) catalyst for the conversion of HMF to DMF was examined by Barta and coworkers [48]. They achieved 90% yield for DMF in concentrated HMF (10 wt%) of cyclopentyl methyl ether (CPME) solution at 493 K under pressurized $H_2$ (20 bar). Catalytic conversion of HMF toward BHMF was also tested with same system, and >95% yield and selectivity were obtained with the CuZn nanopowder catalyst at 393 K and $H_2$ (70 bar) in ethanol solvent.

Recently, Yang and coworkers provided 90% yield of DMF from HMF at 483 K with formic acid in a stainless steel autoclave under self-generated pressure [49]. In this report, a reaction route involving 5-MF instead of BHMF as an intermediate has been ascertained over NiCo/C catalyst. The reaction system proceeded over alloys with non-noble metals would be a futural target for next generation.

## 2.2. Synthesis of DMTHF from HMF hydrogenation

Owing to the ring-hydrogenation step, direct synthesis of DMTHF from HMF (**Scheme 5**) would require different strategies. Yang and Sen applied the water-soluble $RhCl_3$ species with HI acid agent in the transformation of fructose to DMTHF in chlorobenzene/water biphasic reaction system under $H_2$ (2.07 MPa) [50]. This powerful catalytic system served the end product of DMTHF or 2-MTHF from various saccharides such as fructose, glucose, sucrose, inulin, cellulose, and corn stover with significant activity (*ca.* 50–80% yield) at 433 K. Further mechanistic study on HMF conversion [51] revealed that the HI agent behaved as both acid catalyst for the initial dehydration of saccharides to the corresponding furans and as reducing agent in hydrogenolysis step of carbinol group in HMF *via* the formation of 5-iodomethylfurfural intermediate (**Scheme 6**). While the Rh species catalyze not only the C═O hydrogenolysis and C═C hydrogenation but also re-hydrogenation of produced iodine to HI with hydrogen. Overall, such efficient catalytic system, that is, the dehydration/reduction ability of HI combined with the hydrogenation/hydrogenolysis ability of the Rh catalyst, effectively facilitated the conversion of fructose to DMTHF.

Mitra *et al.* carefully surveyed the effects of HMF concentration, Pd loading and kinds of acidic additive in Pd/C-catalyzed hydrogenation/ hydrogenolysis of HMF at 0.21 MPa $H_2$ in

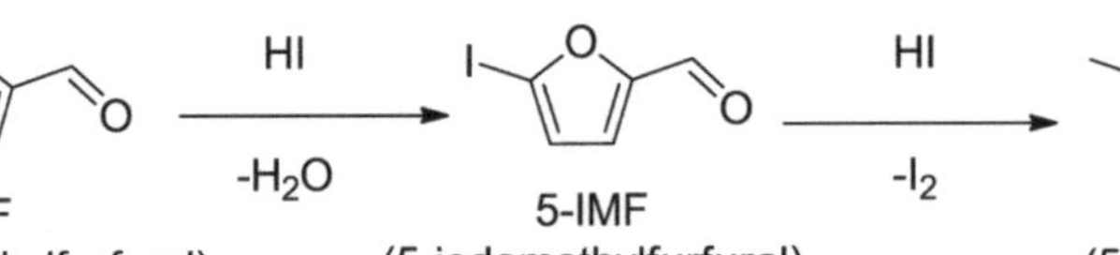


**Scheme 5.** Synthesis of DMTHF from HMF hydrogenation/hydrogenolysis.

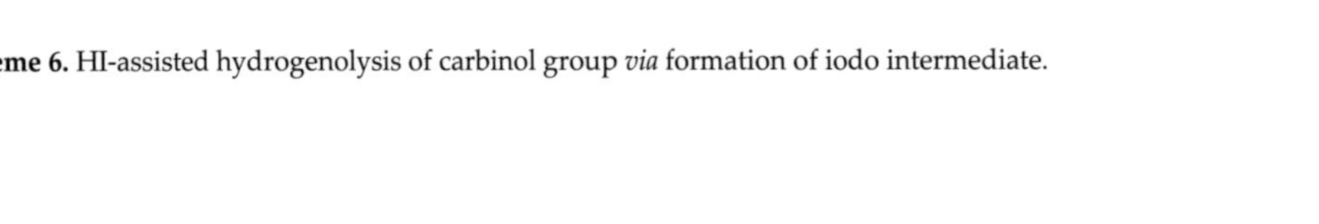


**Scheme 6.** HI-assisted hydrogenolysis of carbinol group *via* formation of iodo intermediate.

dioxane. Under the optimized condition, DMTHF (>95% yield) and DMF (85% yield) was produced in the presence of acetic acid and dimethyldicarbonate, respectively. It is likely that the tendency for ring-hydrogenation against hydrogenation/hydrogenolysis was decreased in higher HMF concentration and amount of catalyst loading [32].

It would be noted that though at higher pressurized $H_2$ condition (70 bar), $Pd/Al_2O_3$ (>99% yield), Pd/C (89% yield), and Ru/C (88% yield) are also found to be a good catalyst for DMTHF production from HMF in ethanol solvent at 393 K, as reported by Bottari *et al.* [48].

### 2.3. Synthesis of DMF from 5-MF or 5-MFA hydrogenation

5-MF based transformations to DMF (**Scheme 7**) has received little attention. The condensed-phase C—O hydrogenolysis of 5-MFA to DMF at 493 K was carried out by the copper chromite ($CuCr_2O_4$·CuO) catalyst pre-reduced with $H_2$ at various temperature (513–633 K) [52]. The authors selected the simplified system in order to examine the relationship of the Cu oxidation state and C—O hydrogenolysis activity which was the one of intermediate steps during hydrogenation/hydrogenolysis of biomass-derived HMF or 5-MF toward DMF. The surface concentration of $Cu^0$ and $Cu^+$ was varied by a reduction temperature, and reduction at 573 K caused the highest activity, an initial rate for the production of DMF and DMTHF divided by BET surface area of catalyst was 108 mmol $m^{-2}$ $min^{-1}$ and the highest concentration of $Cu^0$ and $Cu^+$ sites. It concluded that the $Cu^0$ was primarily responsible for the activity; however, unfortunately, the detailed role of $Cu^+$ could not be ruled out at that time.

### 2.4. Synthesis of 2-MF and 2-MTHF from furfural or FFA hydrogenation

Furfural derived reaction path for 2-MF and 2-MTHF production (**Scheme 8**) has attracted many scientists. The gas-phase selective cleaving the C=O bond outside the furan ring of furfural was performed on $Mo_2C$ [53]. It was considered that the strong interaction between

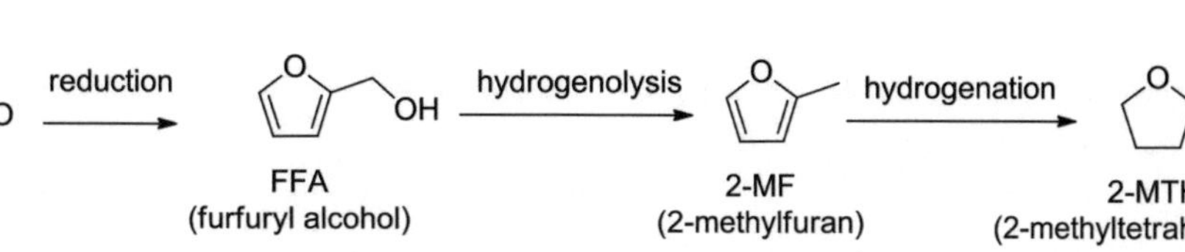


**Scheme 7.** Synthesis of DMF from 5-MF or 5-MFA hydrogenation.




**Scheme 8.** Synthesis of 2-MF and 2-MTHF from furfural or FFA hydrogenation.

C—O bond of furfural in the aldehyde group and $Mo_2C$ surface-assisted selective deoxygenation affording to 60% selectivity for 2-MF production, although the conversion of furfural seems to be low. Zhu *et al.* investigated a fixed-bed reactor test of furfural hydrogenation over a commercial catalyst (Cu/Zn/Al/Ca/Na = 59/33/61/1/1, atomic ratio) [54]. The 87% yield from furfural and 93% yield from FFA were obtained toward 2-MF formation at 523 K. In the same group, recently, further advanced examination on a continuous two-step fixed-bed reactor system over catalysts utilizing a combination of acidic Hβ-zeolite for the former dehydrogenation step of xylose to furfural and $Cu/ZnO/Al_2O_3$ for the latter hydrogenation steps of produced furfural in a mixture of γ-butyrolactone/water as solvent was attempted to the direct conversion of xylose to 2-MF *via* FFA formation [55]. It was observed that the main product can be simply tuned by the hydrogenation temperature, and the 87% yields of FFA and 2-MF was formed at 423 and 463 K under 0.1 MPa $H_2$, respectively. The 2-MF was frequently detected at higher reaction temperature in the FFA synthesis from hydrogenation of furfural with moderate yield (35%) [56].

In the FFA hydrogenation reaction, $Pd/Fe_2O_3$ catalyst yielded 31% for 2-MF under batch conditions at a transfer hydrogenation of FFA in 2-propanol, though the conversion of furfural to 2-MF was scarcely occurred (13% yield) at 453 K [31]. The former value reached to be 76% in a continuous-flow reactor, while the $Pd/TiO_2$ catalyst showed 71 and 88% yield for 2-MF for constant and continuously controlled pressured $H_2$ (0.3 MPa) condition in an autoclave reactor at room temperature (298 K) [57].

In order to avoid the use of high $H_2$ pressure, Zhu *et al.* designed a two-stage packing fixed-bed reactor system with $Cu_2Si_2O_5(OH)_2$ (copper phyllosilicate) in the upper reactor conjugated with $Pd/SiO_2$ catalyst in the bottom reactor [11]. The $Cu_2Si_2O_5(OH)_2$ itself possessed an outstanding hydrogenation ability and gave 84.6% yield for hydrogenation-deoxygenation of furfural toward 2 MF, while $Pd/SiO_2$ catalyst gave a significant hydrogenation performance for the C=C bonds in furan ring, serving 87.7% yield of ring-hydrogenation of 2 MF toward 2 MTHF at 443 K. Accordingly, such a dual solid catalyst system achieved up to 97.1% yield on the direct production for 2-MTHF with 100% conversion of furfural under atmospheric pressure (**Figure 2**). Thereafter, bimetallic CuPd-supported $ZrO_2$ catalyzed hydrogenation by using a batch reactor in the presence of 2-propanol as hydrogen donor was highlighted by other research group [58]. The molecular ratio of Cu/Pd gave significant influences on the yields for 2-MF and 2-MTHF; for example, the highest yield for 2-MF (63.6%) and 2-MTHF (78.8%) was obtained over $Cu_{10}Pd_1/ZrO_2$ and $Cu_{10}Pd_5/ZrO_2$ catalyst, respectively, with 2-propanol (14 mL) at 493 K in 1-mmol scale operation. It was denoted that the reaction mainly proceed through furfural → FFA → 2-MF → 2-MTHF sequence.

To clarify the crucial factor to prevent C—C bonds breaking while effectively breaking the C—O bonds in formyl group in furfural to form 2-MF, suppressing the furan formation, Resasco *et al.* studied the addition of 5-wt% Fe into 2-wt% $Ni/SiO_2$ catalyst with coimpregnation method formed Ni-Fe bimetallic alloys and yielded 39% of MF from furfural and *ca.* 78% from FFA on $NiFe/SiO_2$ catalyst at 523 K in vapor-phase conversion of furfural [59]. Controlled

http://dx.doi.org/10.5772/67805

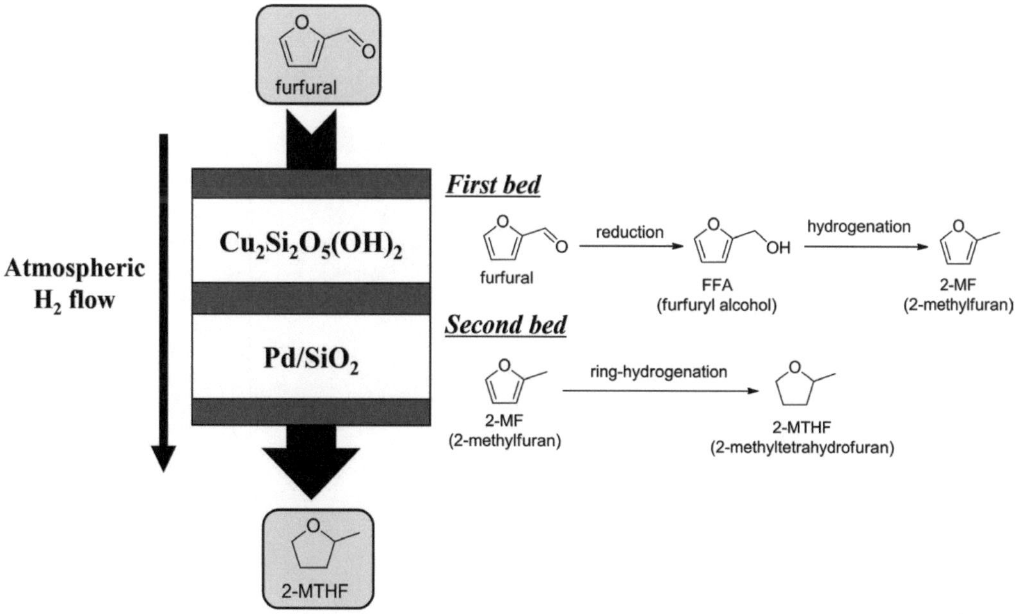


**Figure 2.** Systematic design for one-step conversion of furfural toward 2-MTHF with a dual solid catalyst in a fixed-bed reactor.

experiments with monitoring the yields of products in the conversion of furfural, FFA, and THFA suggested that enhancement of both hydrogenolysis and reduction in decarbonylation activity increased following to increase the Fe content in bimetallic catalyst. Notably, the authors further investigated the reaction with benzyl alcohol, which strongly supported that coexistence of Fe in Ni-Fe/$SiO_2$ catalyst significantly promoted the hydrogenolysis path to toluene rather than the decarbonylation path to benzene (**Figure 3**). The DFT calculation approaches determined that configuration of the adsorbed furfural on NiFe was much differed from Ni, and the significant lengthening of the C—O bond in formyl group caused on bimetallic NiFe surface would be a crucial factor to facilitate the reaction for C—O hydrogenolysis rate observed experimentally.

The same group also has claimed the importance of the differences in the molecular-surface interactions between the aldehydes and active surfaces affording to furan or FFA during furfural hydrogenation over PdCu/$SiO_2$ catalyst in a flow reactor [60]. In the case of PdCu/$SiO_2$, the formation of FFA *via* reduction gradually became dominant in the hydrogenation reaction of furfural in comparison with the formation of furan *via* decarbonylation as function of Cu loadings onto Pd/$SiO_2$ prepared with co-impregnation method; that is, the decarbonylation rate is greatly reduced on PdCu catalyst, but the hydrogenation (reduction) rate is increased. The changes in electron structure of PdCu alloy different from that of pure Pd resulted in a lower extent of electron back-donation to the $\pi^*$ system of the aldehydes and a less stability toward the $\eta^1$(C)-acyl intermediate formation affording to furan, and these increased

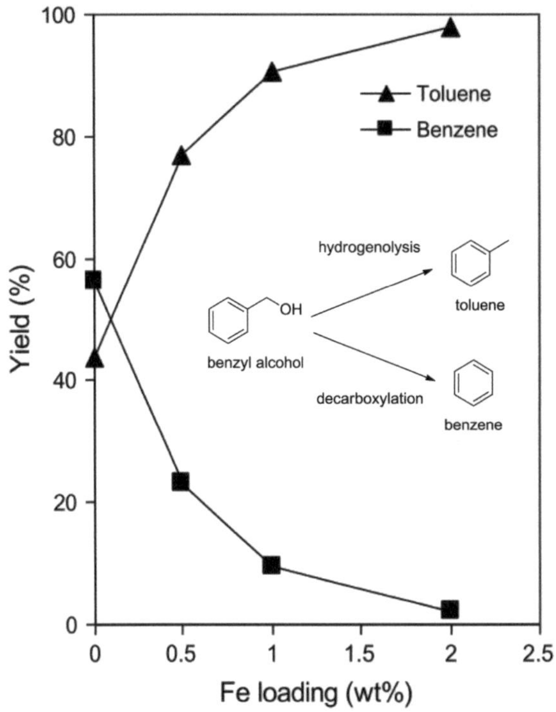


**Figure 3.** Yield of products from the reaction of benzyl alcohol over Ni-Fe bimetallic catalysts as a function of Fe loading. Reaction conditions: W/F = 0.1 h, $H_2$/Feed ratio = 25, 523 K, 1 atm. Reproduced with permission from Elsevier of Ref. [59].

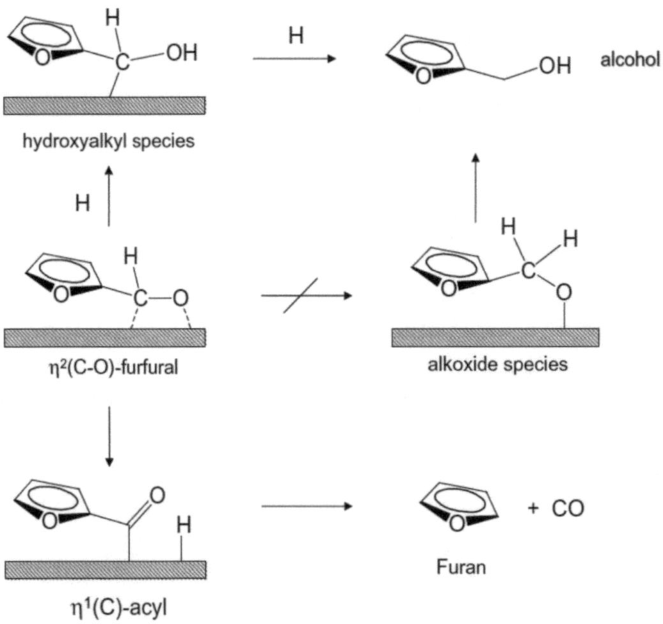


**Figure 4.** An expected mechanism for furfural conversion over Pd catalyst. Reproduced with permission from Elsevier of Ref. [60].

the frequency for FFA production drastically through the hydroxyalkyl intermediate path (**Figure 4**).

Mechanistic studies for understanding and controlling the reactivity for hydrogenation reactions on metal catalysts have been an attractive subject. Advanced review articles for this area would be helpful for further study (e.g., see [61, 62]).

## 3. Conclusions

Due to the high oxygen content in biomass resources, investigations on deoxygenation *via* hydrogenation/hydrogenolysis reaction open up a lot of insights for catalytic and systematic design on biomass upgrading toward biofuel and blender. As introduced in this chapter, bimetallic catalyst and/or utilization of transfer hydrogen donor have been a growing interest instead of conventional system using pressurized $H_2$ and monometalic catalyst. Mechanistic studies with systematic experiments have been a powerful tool to reveal the reaction path and optimize the reaction conditions to provide target compound selectively. As the next research generation, alloying of terminal precious sites with addition of transition metals would have been getting more attentions in hydrogenation/hydrogenolysis of furaldehydes. It needs to be underlined that pyrolysis of biomass resources toward gas, char, and/or biooil is an energetic movement [63–65]. These transformation technologies of biomass would be a key component to fabricate sustainable social design with low carbon emission.

## Author details

Shun Nishimura* and Kohki Ebitani

*Address all correspondence to: s_nishim@jaist.ac.jp

Graduate School of Advanced Science and Technology, Japan Advanced Institute of Science and Technology (JAIST), Nomi, Ishikawa, Japan

## References

[1] World Energy Outlook 2016. France: International Energy Agency (IEA); 2016. ISBN: 978-92-64-26494-6.

[2] 2016 Energy Outlook: BP. Available from: https://www.bp.com/content/dam/bp/pdf/energy-economics/energy-outlook-2016/bp-energy-outlook-2016.pdf [Accessed: 2016-01-12]

[3] Cavani F, Albonetti S, Basile F, Gandini A, editors. Chemicals and Fuels from Bio-Based Building Blocks. Germany: Wiley; 2016. ISBN: 978-3-527-33897-9

[4] Takagaki A, Nishimura S, Ebitani K. Mechanistic studies of solid acids and bases catalyzed clean technologies. In: Wilson K, Lee AF, editors. Heterogeneous Catalysts for

Clean Technology: Spectroscopy, Design, and Monitoring. Germany: Wiley; 2013. pp. 125-171. DOI: 10.1002/9783527658985.ch6

[5] Isikgor FH, Becer CR. Lignocellulosic biomass: a sustainable platform for the production of bio-based chemicals and polymers. Polym. Chem. 2015; **6**: 4497-4559. DOI: 10.1039/c5py00263j

[6] Takagaki A, Nishimura S, Ebitani K. Catalytic transformations of biomass-derived materials into value-added chemicals. Catal. Surv. Asia. 2012; **16**: 164-182. DOI: 10.1007/s10563-012-9142-3

[7] Nishimura S, Ebitani K. Selective oxidation of biomass-derived alcohols with supported metal catalysts. J. Jpn. Petrol. Inst. 2017; **60**: 72-84. DOI: 10.1627/jpi.60.72

[8] Hu L, Zhao G, Hao W, Tang X, Sun Y, Lin L, Liu S. Catalytic conversion of biomass-derived carbohydrates into fuels and chemicals via furanic aldehydes. RSC Adv. 2012; **2**: 11184-11206. DOI: 10.1039/c2ra21811a

[9] Nakagawa Y, Takada K, Tamura M, Tomishige K. Total hydrogenation of furfural and 5-hydroxymethylfurfural over supported Pd-Ir alloy catalyst. ACS Catal. 2014; **4**: 2718-2726. DOI: 10.1021/cs500620b.

[10] Nakagawa Y, Nakazawa H, Watanabe H, Tomishige K. Total hydrogenation of furfural over a silica-supported nickel catalyst prepared by the reduction of a nickel nitrate precursor. ChemCatChem. 2012; **4**: 1791-1797. DOI: 10.1002/cctc.201200218

[11] Dong F, Zhu Y, Ding G, Cui J, Li X, Li Y. One-step conversion of furfural into 2-methyltetrahydrofuran under mild conditions. ChemSusChem. 2015; **8**: 1534-1537. DOI: 10.1002/cssc.201500178

[12] Resasco DE, Sitthisa S, Faria J, Prasomsri T, Ruiz MP. In: Kubicka D, and Kubickova I, editors. Heterogeneous Catalysis in Biomass to Chemicals and Fuels. India: Research Signpost; 2012, pp. 155-188.

[13] Ohyama J, Esaki A, Yamamoto Y, Arai S, Satsuma A. Selective hydrogenation of 2-hydroxymethyl-5-furfural to 2,5-bis(hydroxymethyl)furan over gold sub-nano clusters. RSC Adv. 2013; **3**: 1033-1036. DOI: 10.1039/c2ra22190j

[14] Ohyama J, Hayashi Y, Ueda K, Yamamoto Y, Arai S, Satsuma A. Effect of $FeO_x$ modification of $Al_2O_3$ on its supported Au catalyst for hydrogenation of 5-hydroxymethylfurfural. J. Phys. Chem. C. 2016; **120**: 15129-15136. DOI: 10.1021/acs.jpss.6b01542

[15] Liu Y, Mellmer MA, Alonso DM, Dumesic JA. Effects of water on the copper-catalyzed conversion of hydroxymethylfurfural in tetrahydrofuran. ChemSusChem. 2015; **8**: 3983-3986. DOI: 10.1002/cssc.201501122

[16] Alamilo R, Tucker M, Chia M, Pagan-Torres Y, Dumesic J. The selective hydrogenation of biomas-derived 5-hydroxymethylfurfural using heterogeneous catalyst. Green. Chem. 2012; **14**: 1413-1419. DOI: 10.1039/c2gc35039d

[17] Perret N, Grigoropoulos A, Zanella M, Manning TD, Claridge JB, Rosseinsky MJ. Catalytic response and stability of nickel/alumina for the hydrogenation of 5-hydroxymethylfurfural in water. ChemSusChem. 2016; **8**: 521-531. DOI: 10.1002/cssc.201501225

[18] Nakagawa Y, Tomishige K. Total hydrogenation of furan derivatives over silica-supported Ni-Pd alloy catalyst. Catal. Commun. 2010; **12**: 154-156. DOI: 10.1016/j.catcom.2010.09.003

[19] Zhong S, Daniel R, Xu H, Zhang J, Turner D, Wyszynski ML, Richards P. Combusion and emissions of 2,5-dimethylfuran in a direct-injection spark-ignition engine. Energy Fuels. 2010; **24**: 2891-2899. DOI: 10.1021/ef901575a

[20] Chen G, Shen Y, Zheng Q, Yao M, Zheng Z, Liu H. Experimental study on combustion and emission characteristics of a diesel engine fuelled with 2,5-dimethylfuran-diesel, n-butanol-diesel and gasoline-diesel blends. Energy. 2013; **54**: 333-342. DOI: 10.1016/j.energy.2013.02.069

[21] Qian Y, Zhu L, Wang Y, Lu X. Recent progress in the development of biofuel 2,5-dimethylfuran. Renew. Sustain. Energy Rev. 2015; **41**: 633-646. DOI: 10.1016/j.rser.2014.08.085

[22] Chidambaran M, Bell AT. A two-step approach for the catalytic conversion of glucose to 2,5-dimethylfuran in ionic liquids. Green. Chem. 2010; **12**: 1253-1262. DOI: 10.1039/c004343e

[23] Saha B, Bohn CM, Abu-Omar MM. Zinc-assisted hydrodeoxygenation of biomass-derived 5-hydroxymethylfurfural to 2,5-dimethylfuran. ChemSusChem. 2014; **7**: 3095-3101. DOI: 10.1002/cssc.201402530

[24] Zu Y, Yang P, Wang J, Liu X, Ren J, Lu G, Wang Y. Efficient production of the liquid fuel 2,5-dimethylfuran from 5-hydroxymethylfurfural over Ru/$Co_3O_4$ catalyst. Appl. Catal. B: Environ. 2014; **146**: 244-248. DOI: 10.1016/j.apcatb.2013.04.026

[25] Nagpure AS, Lucas N, Chilukuri SV. Efficient preparation of liquid fuel 2,5-dimethylfuran from biomass-derived 5-hydroxymethylfurfural over Ru-NaY catalyst. ACS Sust. Chem. Eng. 2015; **3**: 2909-2916. DOI: 10.1021/acssuschemeng.5b00857

[26] Hu L, Tang X, Xu J, Wu Z, Lin L, Liu S. Selective transformation of 5-hydroxymethylfurfural into the liquid fuel 2,5-dimethylfuran over carbon-supported ruthenium. Ind. Eng. Chem. Res. 2014; **53**: 3056-3064. DOI: 10.1021/ie404441a

[27] Jae J, Zheng W, Lobo RF, Vlachos DG. Production of dimethylfuran from hydroxymethylfurfural through catalytic transfer hydrogenation with ruthenium supported on carbon. ChemSusChem. 2013; **6**: 1158-1162. DOI: 10.1002/cssc.201300288

[28] Huang YB, Chen MY, Yan L, Guo QX, Fu Y. Nickel-tungsten carbide catalysts for the production of 2,5-dimethylfuran from biomass-derived molecules. ChemSusChem. 2014; **7**: 1068-1070. DOI: 10.1002/cssc.201301356

[29] Chatterjee M, Ishizaka T, Kawanami H. Hydrogenation of 5-hydroxymethylfurfural in supercritical carbon dioxide-water: a turnable approach to dimethylfuran selectivity. Green. Chem. 2014; **16**: 1543-1551. DOI: 10.1039/c3gc42145g

[30] Nagpure AS, Venugopal AK, Lucas N, Manikandan M, Thirmalaiswamy R, Chilukuri, S. Renewable fuels from biomass-derived compounds: Ru-containing hydrotalcites as catalysts for conversion of HMF to 2,5-dimethylfuran. Catal. Sci. Technol. 2015; **5**: 1463-1472. DOI: 10.1039/c4cy01376j

[31] Scholz D, Aelling C, Hermans L. Catalytic transfer hydrogenation/hydrogenolysis for reductive upgrading of furfural and 5-(hydroxymethyl)furfural. ChemSusChem. 2014; **7**: 268-275. DOI: 10.1002/cssc.201300774

[32] Mitra J, Zhou X, Rauchfuss T. Pd-C-catalyzed reactions of HMF: decarbonylation, hydrogenation, and hydrogenolysis. Green Chem. 2015; **17**: 307-313. DOI: 10.1039/c4gc01520g

[33] Thananatthanachon T, Rauchfuss TB. Efficient production of the liquid fuel 2,5-dimethylfuran from fructose using formic acid as a reagent. Angew. Chem. Int. Ed. 2010; **49**: 6616-6618. DOI: 10.1002/anie.201002267

[34] De S, Dutta S, Saha B. One-pot conversions of lignocellulosic and algal biomass into liquid fuels. ChemSusChem. 2012; **5**: 1826-1833. DOI: 10.1002/cssc.201200031

[35] Hansen TS, Barta K, Anastas PT, Ford PC, Riisager A. One-pot reduction of 5-hydroxymethylfurfural via hydrogenation transfer from supercritical methanol. Green. Chem. 2012; **14**: 2457-2461. DOI: 10.1039/c2gc35667h

[36] Hu X, Li CZ. Levulinic esters from the acid-catalysed reactions of sugars and alcohols as part of a bio-refinery. Green. Chem. 2011; **13**: 1676-1679. DOI: 10.1039/C1GC15272F

[37] Luijkx GCA, Huck NPM, van Rantwijk F, Maat L, Bekkum van H. Ether Formation in the hydrogenolysis of hydroxymethylfurfural over palladium catalysts in alcoholic solution. Heterocycles. 2009; **77**: 1037-1044. DOI: 10.3987/COM-08-S(F)81

[38] Alonso DM, Wettstein SG, Dumesic JA. Bimetallic catalysts for upgrading of biomass to fuels and chemicals. Chem. Soc. Rev. 2012; **41**: 8075-8098. DOI: 10.1039/C2CS35188A

[39] Besson M, Gallezot P, Pinel C. Conversion of biomass into chemicals over metal catalysts. Chem. Rev. 2014; **114**: 1827-1870. DOI: 10.1021/cr4002269

[40] Li H, Fang Z, Smith Jr, RL, Yang S. Efficient valorization of biomass to biofuels with bifunctional solid catalytic materials. Prog. Energy Comb. Sci. 2016; **55**: 98-194. DOI: 10.1016/j.pecs.2016.04.004

[41] Nishimura S, Ikeda N, Ebitani K. Selective hydrogenation of biomass-derived 5-hydroxymethylfurfural (HMF) to 2,5-dimethylfuran (DMF) under atmospheric hydrogen pressure over carbon supported PdAu bimetallic catalyst. Catal. Today. 2014; **232**: 89-98. DOI: 10.1016/j.cattod.2013.10.012

[42] Choudhary H, Nishimura S, Ebitani K. Hydrothermal preparation of robust boehmite-supported *N,N*-dimethyldodecylamine *N*-oxide-capped CoPd catalyst for facile utilization of formic acid as a hydrogen source. ChemCatChem. 2015; **7**: 2361-2369. DOI: 10.1002/cctc.201500161

[43] Nishimura S, Yoshida N, Ebitani K. Bimetallic PdCu nanoparticle catalyst supported on hydrotalcite for selective aerobic oxidation of benzyl alcohol. MRS Proc. 2015; **1760**: mrsf14-1760-yy05-32. DOI: 10.1557/opl.2015.58

[44] Gallo JMR, Alonso DM, Mellmer MA, Dumesic JA. Production and upgrading of 5-hydroxymethylfurfural using heterogeneous catalysts and biomass-derived solvents. Green. Chem. 2013; **15**: 85-90. DOI: 10.1039/c2gc36536g

[45] Román-Leshkov Y, Barrett CJ, Liu ZY, Dumesic JA. Production of dimethylfuran for liquid fuels from biomass-derived carbohydrates. Nature. 2007; **447**: 982-986. DOI: 10.1038/nature05923

[46] Yu L, Chen J, Zheng J, Ye L, Lin H, Yuan Y. Robust and recyclable nanoprecious bimetallic nanoparticles on carbon nanotubes for the hydrogenation and hydrogenolysis of 5-hydroxymethylfurfural. ChemCatChem. 2015; **7**: 1701-1707. DOI: 10.1002/cctc.201500097

[47] Wang GH, Hilget J, Richter FH, Wang F, Bongard HJ, Splithoff B, Weidenthaler C, Schuth F. Platinum-cobalt bimetallic nanoparticles in hollow carbon nanospheres for hydrogenolysis of 5-hydroxymeyhylfurfural. Nat. Mater. 2014; **13**: 293-300. DOI: 10.1038/NMAT3872

[48] Bottari G, Kumalaputri AJ, Krawczyk KK, Feringa BL, Heeres HJ, Barta K. Copper–zinc alloy nanopowder: a robust precious-metal-free catalyst for the conversion of 5-hydroxymethylfurfuraln. ChemSusChem. 2015; **8**: 1323-1327. DOI: 10.1002/cssc.201403453

[49] Yang P, Xia Q, Liu X, Wang Y. Catalytic transfer hydrogenation/hydrogenolysis of 5-hydroxymethylfurfural to 2,5-dimethylfuran over Ni-Co/C catalyst. Fuel. 2017; **187**: 159-166. DOI: 10.1016/j.fuel.2016.09.026

[50] Yang W, Sen A. One-step catalytic transformation of carbohydrates and cellulosic biomass to 2,5-dimethyltetrahydrofuran for liquid fuels. ChemSusChem. 2010; **3**: 597-603. DOI: 10.1002/cssc.200900285

[51] Grochowski MR, Yang W, Sen A. Mechanistic study of a one-step catalytic conversion of fructose to 2,5-dimethyltetrahydrofuran. Chem. Eur. J. 2012; **18**: 12363-12371. DOI: 10.1002/chem.201201522

[52] Deutsch KL, Shanks BH. Active species of copper chromite catalyst in C-O hydrogenolysis of 5-methylfurfuryl alcohol. J. Catal. 2012; **285**: 235-241. DOI: 10.1016/j.jcat.2011.09.030

[53] Xiong K, Lee WS, Bhan A, Chen JG. Molybdenum carbide as a highly selective deoxygenation catalyst for converting furfural to 2-methylfuran. ChemSusChem. 2014; **7**: 2146-2149. DOI: 10.1002/cssc.201402033

[54] Zheng HY, Zhu YL, Teng BT, Bai ZQ, Zhang CH, Xiang HW, Li YW. Towards understanding the reaction pathway in vapor phase hydrogenation of furfural to 2-methylfuran. J. Mol. Catal. A: Chem. 2006; **246**: 18-23. DOI: 10.1016/j.molcata.2005.10.003

[55] Cui J, Tan J, Cui X, Zhu Y, Deng T, Ding G, Li Y. Conversion of xylose to furfuryl alcohol and 2-methylfuran in a continuous fixed-bed reactor. ChemSusChem. 2016; **9**: 1259-1262. DOI: 10.1002/cssc.201600116

[56] Panagiotopoulou P, Martin N, Vlachos DG. Liquid-phase catalytic transfer hydrogenation of furfural over homogeneous Lewis acid-Ru/C ctalysts. ChemSusChem. 2015; **8**: 2046-2054. DOI: 10.1002/cssc.201500212

[57] Iqbal S, Liu X, Aldosari OF, Miedziak PJ, Edwards JK, Brett GL, Akram A, King GM, Davies TE, Morgan DJ, Knight DK, Hutchings GJ. Conversion of furfuryl alcohol into 2-methylfuran at room temperature using Pd/$TiO_2$ catalyst. Catal. Sci. Technol. 2014; **4**: 2280-2286. DOI: 10.1039/c4cy00184b

[58] Chang X, Liu A, Cai B, Luo J, Pan H, Huang Y. Catalytic transfer hydrogenation of furfural to 2-methylfuran and 2-methyltetrahydrofuran over bimetallic copper-palladium catalysts. ChemSusChem. 2016; **9**: 3330-3337. DOI: 10.1002/cssc.201601122

[59] Sitthisa S, An W, Resasco DE. Selective conversion of furfural to methylfuran over silica-supported Ni-Fe bimetallic catalysts. J. Catal. 2011; **284**: 90-101. DOI: 10.1016/k/jcat.2011.09.005

[60] Sitthisa, S, Pham T, Prasomsri T, Sooknoi T, Mallinson RG, Resasco DE. Conversion of furfural and 2-methylpentanal on Pd/$SiO_2$ and Pd-Cu/$SiO_2$ catalysts. J. Catal. 2011; **280**: 17-27. DOI: 10.1016/j.jcat.2011.02.006

[61] Gilkey MJ, Xu B. Heterogeneous catalytic transfer hydrogenation as an effective pathway in biomass upgrading. ACS Catal. 2016; **6**: 1420-1436. DOI: 10.1021/acscatal.5b02171

[62] Medlin JW. Understanding and controlling reactivity of unsaturated oxygenates and polyols on metal catalysts. ACS Catal. 2011; **1**: 1284-1297. DOI: 10.1021/cs200336r

[63] Bridgwater AV, Peacocl GVC. Fast pyrolysis processes for biomass. Renew. Sust. Energy Rev. 2000; **4**: 1-73. DOI: 10.1016/S1364-0321(99)00007-6

[64] Mohan D, Pittman CU, Jr., Steele PH. Pyrolysis of wood/biomass for bio-oil: a critical review. Energy Fuel. 2006; **20**: 848-889. DOI: 10.1021/ef0502397

[65] Nishimura S, Miyazato A, Ebitani K. Properties of bio-oil generated by a pyrolysis of forest cedar residuals with the movable Auger-type reactor. AIP Conf. Proc. 2016; **1709**: 020026. DOI: 10.1063/1.4941225

Section 3

# Green Synthesis and Catalysis

Chapter 4

# Green Synthesis of Oligomer Calixarenes

Ratnaningsih Eko Sardjono and Rahmi Rachmawati

Additional information is available at the end of the chapter

http://dx.doi.org/10.5772/67804

**Abstract**

The synthesis of calixarenes can be conventionally done by heating at high temperature for a few hours and using various solvents in large quantities. The greener synthesis can be done with microwave-assisted synthesis and the solvent-free method, where both of these methods can reduce reaction time, energy use, solvent, and waste, with a higher percentage yield than that from the conventional synthesis method, making the synthesis of cyclic oligomer calixarenes and their derivatives more environmentally friendly.

**Keywords:** calixarenes, cyclic oligomer, green synthesis, microwave, solvent-free

## 1. Introduction

A calixarene is a cup-shaped supramolecule. The name is derived from the word "calix," which means a cup. Synthesis of calixarenes becomes very important because these molecules are widely used in various fields, especially in its use in a guest-host system molecule. The molecule was first reported in 1872 by Baeyer as a reaction product of aldehyde-phenol condensation. The same product can also be formed from benzaldehyde-pyrogallol (benzene-1,2,3-triol) condensation reaction. The structure of phenols used in this reaction has reactivity at ortho and para positions so that it can form a polymer cross-linked at three positions (two orthos and one para). In 1942, Zinke and Ziegler used para-substituted phenols so that the phenols could only react at the two ortho positions, thereby reducing the possibility of cross-link formation. By reacting *p-tert*-butylphenol with formaldehyde under a basic condition at 100–220°C, a waxy brown paste was produced, which was then recrystallized and formed a crystal decomposed at above 300°C. The results of the analysis showed that the product had a molecular weight that was in agreement with a cyclic tetrameric structure [1]. Similarly, Cornforth reacted *p-tert*-butylphenol with formaldehyde. However, the reaction resulted in two crystal products that were rather soluble in water with high but nonidentical melting points. Likewise, when *p-tert*-butylphenol was replaced with *p*-1,1,3,3-tetramethylbutylphenol, two products with different melting points were obtained. The results of X-ray showed that the four condensation products had a cyclic tetrameric structure. The different melting points, according to Cornforth, indicated that the two compounds are diastereoisomers [1, 2].

Calixarene synthesis can be done through a noncomplicated process. The compound group is commonly synthesized from 4-alkylphenol, especially 4-*t*-buthylphenol, through a one-stage reaction with formaldehyde and NaOH or KOH as the base. The making of calixarenes, known as a base-induced process, results in calixarenes that may vary in their aromatic components. The reaction product between phenols or phenol derivatives and aldehydes is also known as calixarene, while the reaction product between resorcinol or resorcinol derivatives and aldehydes is known as calixresorcinarene. In addition, there are several products that use aromatic compounds in the forms of para-alkoxy-substituted benzyl alcohol derivatives, such as *p*-alkoxybenzyl alcohol, that when reacted with the aid of $AlCl_3$ as a catalyst will result in non-hydroxylated calixarenes [3]. In general, these reactions are illustrated in **Figure 1**.

The basic structure of calixarenes consists of phenolic units or phenolic derivatives which are repeated and interconnected with Ar-C-Ar bridge to form a hollow cycle. The part with a broader hollow side is called the upper rim, and the narrow hollow part is called the "lower rim" as shown in **Figure 2** [4, 5].

The various conformations of calixarenes are caused by the free rotation of sigma bonds at methylene, Ar-C-Ar, bridges [6]. For example, calix[4]resorcinarenes have four prochiral centers on carbon atoms at the methylene bridges, causing the compound to form four different diastereoisomers. The relative configurations of the four R substituents at the methylene bridges can form a cis for all R substituents (rccc) or a cis-cis-trans to one R (rcct), cis-trans-trans (rctt), and trans-cis-trans (rtct). The configurations are determined based on their configurations to one of the R substituents on the prochiral center in a clockwise direction [4, 7, 8]. **Figure 3** illustrates the relative configurations of the four R substituents.

In reality, the cyclic tetrameric ring of a calixarene is not a planar molecule and can adopt five conformations, namely, "crown" or $C_{4v}$, "boat" or $C_{2v}$, "chair" or $C_{2h}$, "diamond" or $C_s$, and "saddle" or $S_4$ [2, 8]. The five conformations of calixarene is shown in **Figure 4**. In fact, the rccc isomer where all R groups in the same position, so far only found with the macrocyclic ring adopted the "crown" or $C_{4v}$ and "boat" or $C_{2v}$ conformations. This is caused by the single bond with a tetrahedral shape of R groups so that it can rotate freely. The rccc isomers can form conformation "boat" or $C_{2v}$ with the fourth R possesses an all axial and all-cis configuration, while the macrocyclic ring in a boat conformation. In this case, the fourth R leads into the annulus of the calixarene (endo). It was also reinforced by the result of a condensation reaction between resorcinol and 4-bromobenzaldehyde with an acid catalyst, where produced the rccc isomer of calix[4]resorcinarene in "boat" or $C_{2v}$ conformation [7]. The rctt isomer is only found in "chair" or $C_{2h}$ conformation. In the calculation of energy, the rccc isomer in the crown conformation ($C_{4v}$) is favored over the rctt isomer in the chair conformation ($C_{2h}$).

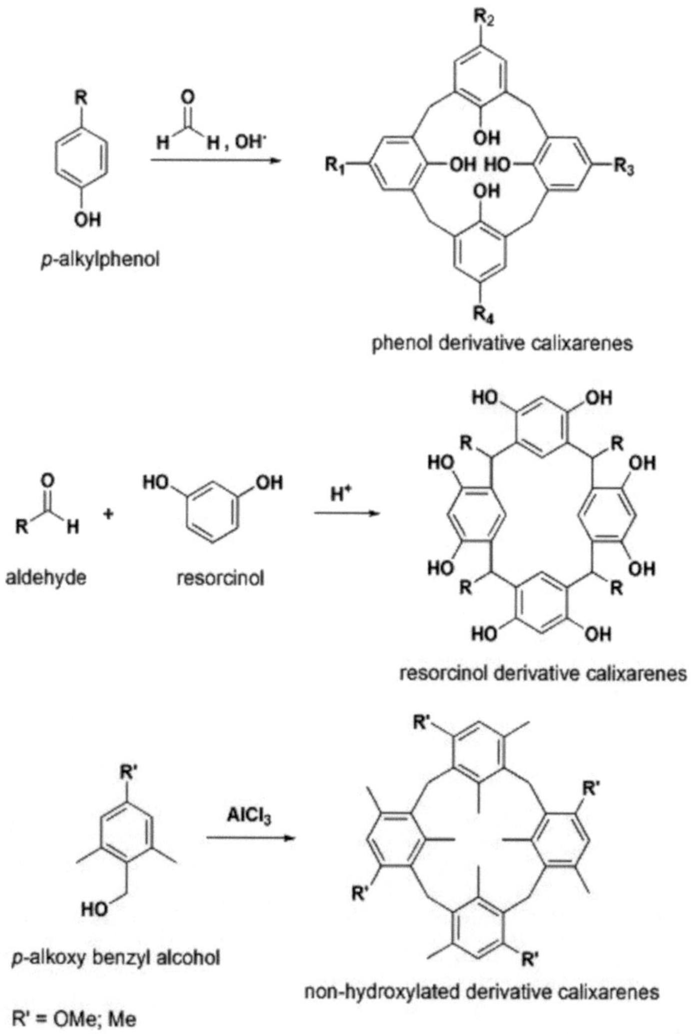


**Figure 1.** Basic synthesis of calixarene.

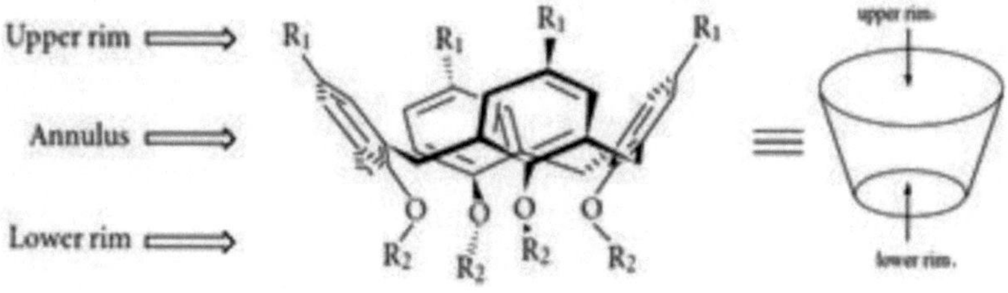


**Figure 2.** The "upper rim" and "lower rim" of calixarenes.

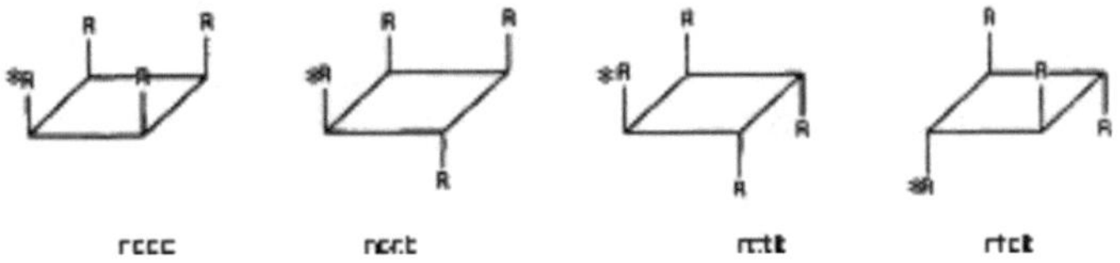


**Figure 3.** R relative configurations at the methylene bridge.

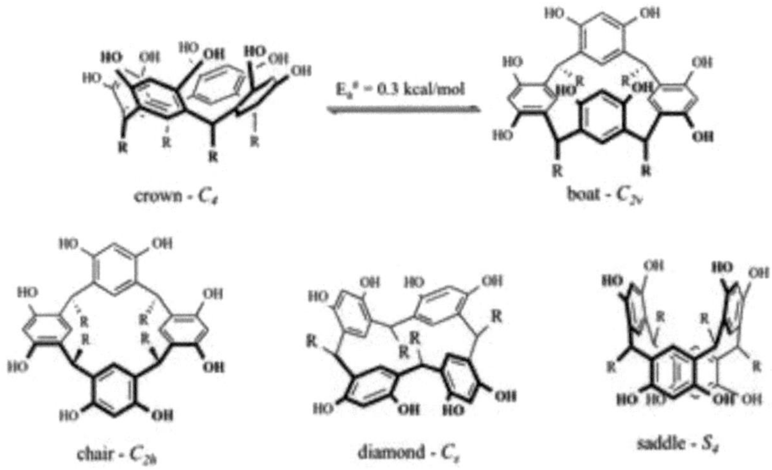


**Figure 4.** Calix[4]arene conformations.

Nevertheless, the rccc in boat conformation ($C_{2v}$) has similar energy to the rctt in chair conformation [51]. The rcct isomer adopts the "diamond" or $C_s$ conformation [4].

Based on the projection of aryl group on the cyclic tetramer structure of calixarenes, four different conformations can be formed depending on the forms of aryl cluster projection, which can be upward (u) or downward (d), relative to the plane. The four conformations are a cone, partial cone, 1,2-alternate, and 1,3-alternate conformations (**Figure 5**), each with different thermodynamic stabilities [1, 5].

In their development, calixarenes have been successfully synthesized not only in the form of the tetramer but also hexamer and octamer. Therefore, the naming of calixarenes is developed by inserting a number inside brackets, showing the number of monomers they possess. For instance, a calixarene containing eight monomers will be written as calix[8]arene. Furthermore, to show the types of phenol used in the synthesis, para substituents are also included in the name. The cyclic hexamer of *p-tert*-butylphenol, for example, is named *p-tert*-butylcalix[6]arene. For resorcinol-derived calixarenes, the name is calix[n]resorcinarene, while the substituents of the methylene carbon are shown by the prefix of "C-substituent." For example, the reaction yield of resorcinol with acetaldehyde is named C-methylcalix[4]resorcinarene, indicating the presence of methyl substituents in the methylene carbon. For a more systematic

http://dx.doi.org/10.5772/67804

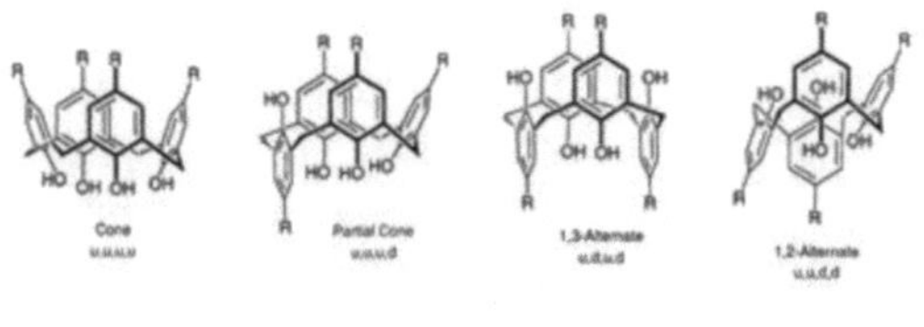


**Figure 5.** Conformation based on the projection of aryl group in tetramer structure of calixarenes.

naming, especially for scientific publication, the basic name of calix[n]arene is maintained, whereas the substituent identity is shown based on its position through the number, such as shown in **Figure 6** [9].

Calixarenes have been broadly used, among others, as liquid crystal materials [10–12], sensors [13, 14], catalysts [15, 16], the stationary phase of chromatography [17], host molecules [18–25], or even semiconductors [26]. As a role in the guest-host system molecule, calix[4]arene and 4-*tert*-butylcalix[6]arene can be the host molecules for compounds such as trifluoromethyl-benzene contained in pesticides [24]. In addition, some calixarene derivatives can also be the host for the dye molecules such as orange I (OI), methylene blue (MB), neutral red (NR), and brilliant green (BG) [18]. This finding makes calixarenes as a promising candidate as a host molecule to be used in chemical sensors to measure neutral organic molecules. In addition to being the host molecules for neutral compounds, calixarene and its derivatives were also proven to be a host for the molecular ion compounds such as heavy metal ions [19, 20]. It is no wonder that calix[n]arenes have been greatly synthesized and yielded various oligomers with n = 4–20 according to their purposes. Various modifications of functional clusters have also been developed in order to yield the desired molecular structures, for example, the modification of the aryl group [27] or the aldehyde. The modification can be in the form of the addition of a functional cluster such as alkyl cluster and hydroxyl cluster or the elimination of a functional cluster in the starting materials used or further modification of the calixarene molecules formed [19]. Some of the modifications make the calixarene structure more rigid,

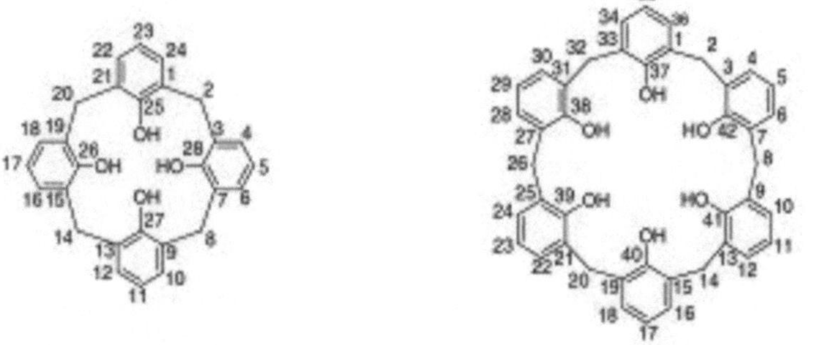

calix[4]arene-25,26,27,28-tetrol          calix[6]arene-37,38,39,40,41,42-hexol

**Figure 6.** Calixarene numbering system.

thereby forming the expected conformations [28]. To cite an example, the addition of ketone group into aryl group was done to increase the affinity of calixarene compound to the complex formation of alkali cation [29]. The modification can also be made by adjusting the hollow size formed in calix[n]arenes to match the size of a certain molecule both by modifying the functional cluster and adding other components [2, 30].

## 2. Synthesis of oligomer calixarenes

### 2.1. Conventional synthesis procedures

Various calixarene synthesis procedures and their derivations have been largely developed, for example, the procedure developed by Zinke-Ziegler, where the synthesis of calixarene started by mixing *p-tert*-butylphenol, 37% formaldehyde, and NaOH heated at 50–55°C for 45 h. To remove water content, the reaction compound was reheated at 110–120°C for 2 h. In the purification process, the yellow solid yielded was subsequently heated in linseed oil up to 200°C for several hours, and the process yielded tetramers. The modification to this procedure was made by Cornforth by directly heating the reaction compound (*p-tert*-butylphenol, formaldehyde, and NaOH) at 110–120°C for 2 h, thereby yielding a very thick product which was then refluxed with diphenyl ether for 2 h. Once the compound was cold, the product was separated with infiltration and recrystallized to produce white crystals with a melting point of 342–344°C for 50%. The procedure was later on known as the Zinke-Cornforth procedure [1, 2, 31]. The condensation of *p-tert*-butylphenol with formaldehyde under basic conditions (NaOH, KOH) can yield cyclic tetramers, hexamers, and even octamers with a percentage of 85–63%. **Figure 7** illustrates the reaction mechanisms of calix[n]arene formation under basic conditions.

The right mechanism to transform linear oligomer into a cyclic oligomer, according to Gutsche [1], can be predicted from the formation of hemicalixarene combined with a pair of linear oligomers. Under appropriate conditions, two linear methyl-hydroxylated tetramers combine to form hemicalix[8]arene, followed by the release of water to form calix[8]arene [1]. This process can also be seen in the synthesis of *p*-chlorocalix[4]arene, which is a "2 + 2" condensation product of *p,p*'-dichlorodiphenylmethane, such as illustrated in **Figure 8**.

Similarly, calix[6]arene formation is postulated to be the result of a combination of two linear trimers. Based on this postulation, the form of dianion hemicalix[6]arene can be the cause of the large amounts of bases required for hemicalix[6]arene synthesis. The following reaction in **Figure 9**, for example, adopts a "3 + 3" condensation that involves two trimers to obtain 9% calix[6]arene [32].

Gutsche synthesized *p-tert*-butylcalix[4]arene with the Zinke-Cornforth procedure to see the effect of basic conditions on the cyclic oligomer produced. The yellow thick paste ("precursor") formed after heating the compound for 1.5–2 h at 110–120°C was diluted with chloroform and washed with aqueous HCl and water several times to neutralize the base. The results of Gutsche's research showed that when the precursor containing a base of 0.03–0.04

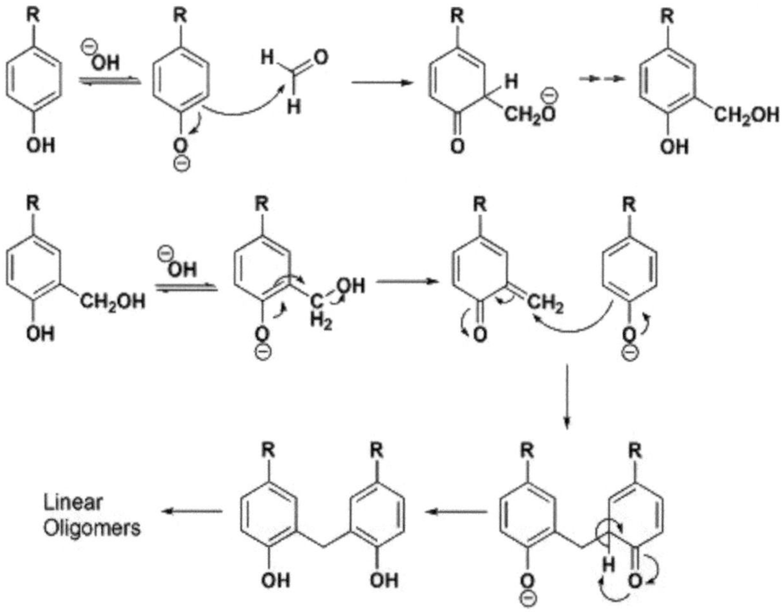


**Figure 7.** Reaction mechanisms of calix[n]arene formation under basic conditions.

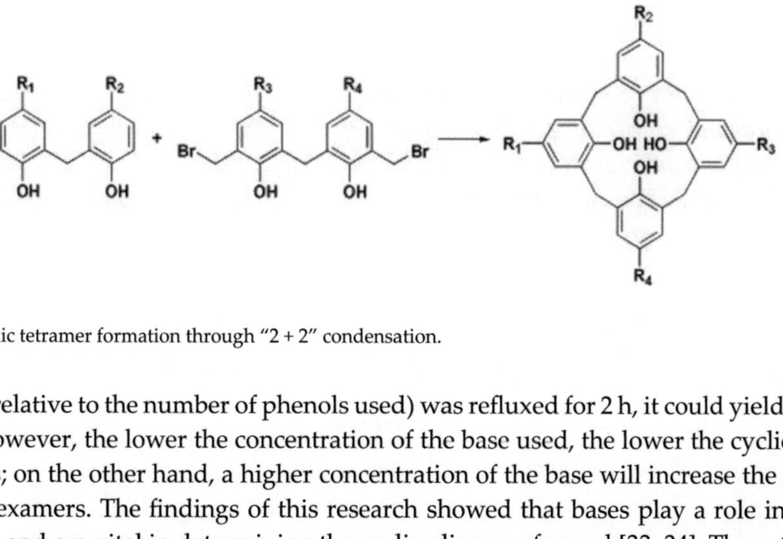


**Figure 8.** Cyclic tetramer formation through "2 + 2" condensation.

or equiv. (relative to the number of phenols used) was refluxed for 2 h, it could yield cyclic tetramers. However, the lower the concentration of the base used, the lower the cyclic tetramer formations; on the other hand, a higher concentration of the base will increase the formation of cyclic hexamers. The findings of this research showed that bases play a role in inducing cyclization and are vital in determining the cyclic oligomer formed [33, 34]. The cations used as the base catalyst can play the role of a "template" which determines the size or number of monomers forming oligomers.

A large cyclic oligomer enables the formation of a larger hollow, and this can be used to absorb heavy metals, as can be found in the synthesis of calix[6]arene involving *p-tert*-butylphenol, paraformaldehyde, and bases (KOH) with a proportion of 0.45 mole relative to the *p-tert*-butylphenol used. The reaction happened for 4 h by refluxing it in *p*-xylene. The compound

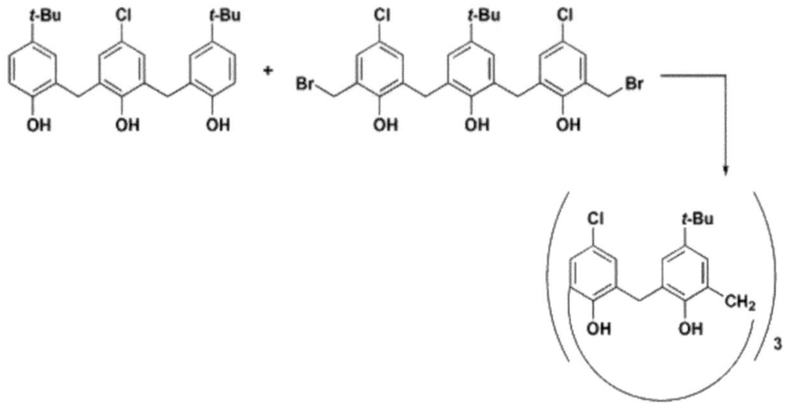


**Figure 9.** Cyclic hexamer formation through "3 + 3" condensation.

was separated with liquid-liquid extraction using chloroform. The organic phase obtained was then evaporated and recrystallized with methanol to yield 65.47% white solids with a melting point of 370–372°C. The analysis of the product showed that it is a cyclic hexamer of calix[6]arene [35].

Condensation of resorcinol with formaldehyde under acidic conditions formed a linier or cross-linked polymer between formaldehyde with resorcinol. This reaction also did not produce calixresorcinarene under alkaline conditions. It is not surprising since resorcinol and formaldehyde are highly reactive so that the reaction becomes uncontrollable and does not produce a cyclic oligomer. Base conditions can be applied to the synthesis of calixresorcinarene from 2-substituted resorcinols and formaldehyde. Nevertheless, resorcinol used should be substituted electron-withdrawing group at position 2 as shown in **Figure 10**. Electron-withdrawing group at position 2 will reduce the reactivity of resorcinol

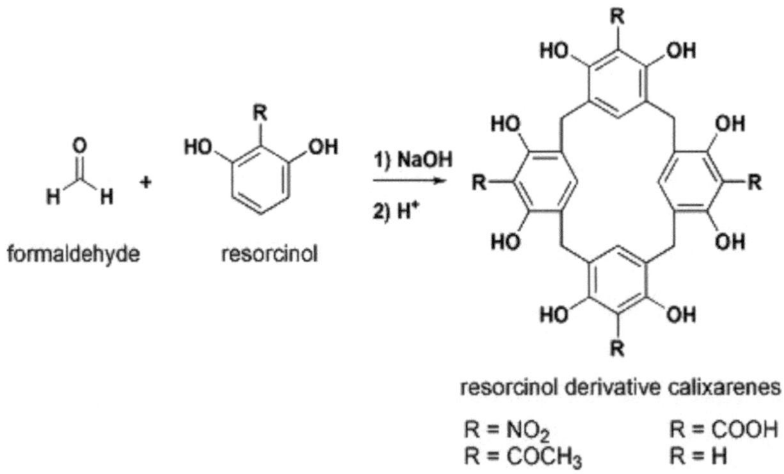


**Figure 10.** Synthesis of resorcinarenes from 2-substituted resorcinols and formaldehyde in basic conditions.

so that the reaction will be controlled and it will be suitably carried out under alkaline conditions. The presence of electron-withdrawing substituents such as nitro group ($-NO_2$) facilitates condensation reactions that occur by directing the reaction site at the meta position of the nitro group. The reaction rate decreased from electron-withdrawing substituent nitro > acetyl > carboxylate > H due to the ability of the groups to deactivate the resorcinol. In contrast, the reaction of resorcinol substituted electron-donating group such as 2-methylresorcinol or phloroglucinol (=benzene-1,3,5-triol) with formaldehyde under base conditions did not produce a derivative of calixresorcinarene. Therefore, the acidic conditions are more suitable when used resorcinol substituted electron-donating group in position 2. The condensation reaction between formaldehyde and 2-nitroresorcinol and its analogs seems to take place in the irreversible reaction, so it is unknown whether to form a tetramer, pentamer, or hexamer [36].

Besides adding bases, the synthesis of calix[n]arenes can also be done under acidic conditions. One such example is found in the procedure of calix[n]resorcinarene synthesis developed by Niederl-Hogberg, in which resorcinol was used as an aromatic unit in the reaction. Resorcinol is more reactive compared to phenol. Resorcinol was reacted with acetaldehyde by adding hydrochloric acid. The reaction took place at 80°C for 16 h and yielded C-methylcalix[4]resorcinarenes for 75%. The reaction follows the following mechanism which is illustrated in **Figure 11**.

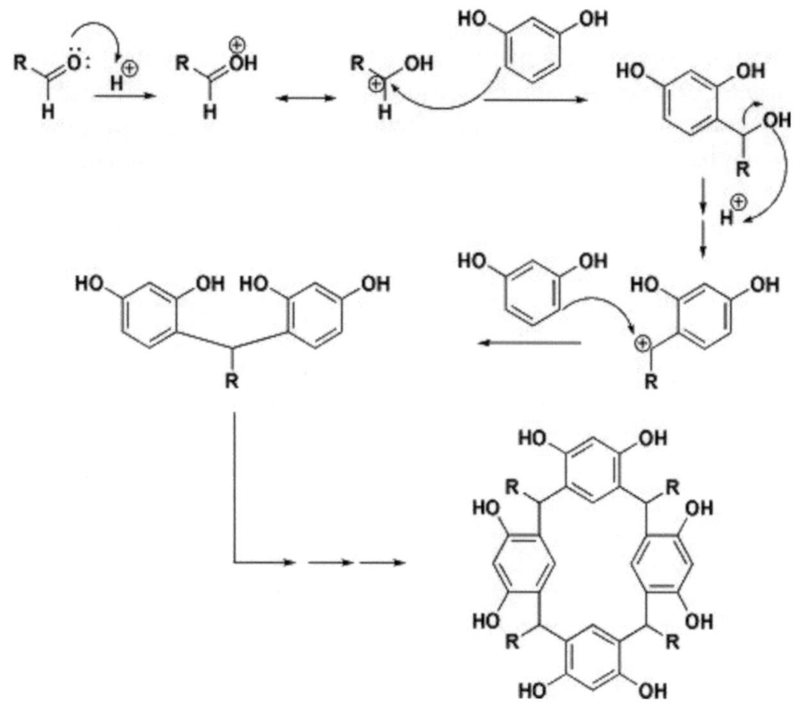


**Figure 11.** Reaction mechanism of calix[n]resorcinarene formation under acidic conditions.

In the condensation reaction catalyzed by acids, kinetic and thermodynamic factors compete with each other. The kinetic factor controls the oligomer rate produced, while the thermodynamic factor controls the cyclic structure formation. In this reaction, controlling the thermodynamic factor will be very determining for the formation of a small cyclic oligomer, such as the formation of calix[4]resorcinarene. To control the thermodynamic factor, less reactive aldehydes can be used to slow down the reaction rate. If the aldehyde used is very reactive (such as formaldehyde), the reaction will happen very quickly so that the kinetic factor will dominate and the majority of the products formed will be polymers with linear chains. A modified aldehyde was used in the synthesis reaction of methoxyphenylcalix[4]resorcinarene from resorcinol with *p*-anisaldehyde with the addition of HCl. The reaction happened employing reflux for 30 h, yielding 4-methoxyphenylcalix[4]resorcinarene as much as 91.54% [37].

The synthesis of calix[4]resorcinarene from resorcinol with acetaldehyde was done by Petrova et al. in 2012. The reaction was made by varying the amounts of hydrochloric acid used. The reaction occurred at 75°C for 5–24 h. The reaction product was dependent upon the amounts of hydrochloric acid used. Increasing the amount of acid can cause the formation of linear oligomers. The amount of acid catalyst effective in this reaction was 4% of the resorcinol. In addition, the polarity of the solvents used affected the composition of the yielded product. An increase in the polarity of the solvent caused an increase in the oligo cyclization. Calix[4]resorcinarene with rccc conformation was quantitatively formed very well (76.89%) when 30% water-ethanol solvent was used [38].

Calix[4]resorcinarenes synthesized through a condensation reaction between resorcinol and an aliphatic or aromatic aldehyde under acidic conditions yield products in the form of cyclic tetramer calixresorcinarenes that theoretically can contain four diastereoisomer products. However, in reality, only three have been identified to be formed by the reaction with an aliphatic aldehyde and two from the reaction of an aromatic aldehyde with resorcinol. The products with crown conformation with all side chains in the axial position (rccc) are more desirable and stable thermodynamically. Hogberg reacted resorcinol with acetaldehyde by adding hydrochloric acid. The reaction occurred at 75°C for 1 h and yielded sediment that was subsequently acetylated. The acetylated product was in the form of two isomeric cyclic tetramers, one of which formed cis (rccc) conformation of up to 47%, while the other product formed rctt conformation with a percentage yield of 13%. This indicated that products with all side chains in the axial position (rccc) are more desirable and stable thermodynamically [39]. The same result can be obtained by selectively increasing the alcohol proportion in the solvent mixtures used and increasing the reaction temperatures and time [39, 40].

Several synthesis methods of calix[n]arenes and their derivatives have been largely developed to produce high yields. However, the methods require heating at high temperatures and a long reaction time. This requirement is disadvantageous, ultimately when the synthesis is done on a large scale. The required solvents will be in a very large amount, and the reaction time needed will be much longer. This makes the synthesis of calix[n]arenes and their derivatives not environmentally friendly.

### 2.2. Green synthesis procedures

Various approaches to calixarenes and their derivatives have been attempted, one of which is by implementing the concept of "green chemistry." The synthesis of calixarenes and their derivatives requires heating at high temperatures for quite a long time. The heating is usually done using conventional heating, where heat transfer occurs conductively. The method is relatively slow and inefficient to transfer energy into the reaction system, as it depends on the thermal conductivity of the materials used; consequently, the temperature of the reaction vessel will be greater than that in the reaction compound. Irradiation with a microwave can produce more efficient internal heating because microwave energy penetrates into the containers so that it can directly hit the reaction compound, and energy transfer becomes more efficient. The phenomenon will depend on the ability of the molecules (solvents or reagents) to absorb microwave energy and change it into heat. The thermal energy is obtained when the polar molecules try to align themselves with an electric field resulted from microwave irradiation. The amount of heat produced depends on the dielectric properties of the molecules. Dielectric constants indicate the ability of the molecules to get polarized by the electric field. The electric field received by the molecules causes reorientation of the dipolar molecules which in turn increases polarity and reactivity by reducing the activation energy. Consequently, the reaction can happen more quickly [41–43].

Organic synthesis using microwave technology has been largely employed, as it can accelerate the reaction time as a consequence of an increase in heating that cannot be achieved with conventional heating [44]. This is what happened in Baozhi's [45] synthesis of *p*-methylcalix[6] arene. A reaction compound consisting of formaldehyde, *p*-methylphenol, and KOH was heated with microwave for 2–8 min (output power at 100%). The required reaction time at this stage was faster than that when using conventional heating, which could take up to 2 h. After solids were formed, reflux was done for 40 min using dimethylbenzene. When cooled down, white crystals with a melting point of 361–362°C were yielded for up to 81.7%. The synthesis of *p-tert*-butylcalix[8]arene was done with the same procedure using the microwave as well, in which the compound was radiated for 4–8 min. To yield a larger cyclic reaction product, the amounts of bases used were increased with a longer reflux time. As previously discussed, bases and reflux time play a role in cyclic formation [45].

The same phenomenon was observed when calixarene synthesis was done using resorcinol, which is known to be more reactive than phenol. The reaction product in the form of cyclic tetramer calix[n]resorcinarene was obtained after the compound was heated with conventional heating for 5–10 h, with a reaction yield of around 60–90%. However, the reaction could be faster when the heating method was changed with microwave. One reaction only required 3–5 min with a greater reaction percentage yield, namely, around 80–90% [46]. This finding was put to use in the synthesis of thiacalix[n]arene with n = 4–8 [47].

Aldehyde modification can also be done by using the abundant natural products such as vanillin that can be derived from vanillin plant (*Vanilla planifolia*), cinnamaldehyde from cinnamon oil, and anisaldehyde from Indonesian fennel oil. The synthesis procedures using these materials have been developed for the synthesis of calix[n]arenes and their derivatives with "green chemistry" approach. The three compounds were reacted with resorcinol to yield cyclic tetramers, calix[4]resorcinarenes. As previously discussed, the synthesis reaction of calix[4]resorcinarenes can occur for 15–30 h using conventional heating. Sardjono et al. [48] developed a microwave-assisted synthesis procedure for calix[4]resorcinarenes, where resorcinol was reacted to vanillin or cinnamaldehyde. The calix[4]resorcinarenes derivatives yielded could be up to 97.9% in 8 min with a ratio of resorcinol to vanillin or cinnamaldehyde was 1:1, and the microwave power used 332 W. Another calix[4]resorcinarenes derivatives, namely C-anisalcalix[4]resorcinarene could also be yielded for up to 99.5% in 5 min from the reaction of resorcinol with anisaldehyde at the ratio of 1:1.2 and 264 W microwave power [48]. The three reactions of resorcinol with various aldehydes to form calix[4]resorcinarenes is illustrated in **Figure 12**.

Vanillin in the synthesis of calix[4]resorcinarene derivatives was also employed by Nakajima and Kobayashi [20], although they still used the conventional heating method (without a microwave). The reaction compound comprised resorcinol and aldehyde (vanillin, syringaldehyde, or *p*-hydroxybenzaldehyde) with a mole ratio of 1:1 and hydrochloric acid heated at 70°C and stirred for 12 h. The reaction yield was then washed with hot water and purified using methanol and ethanol. The third reaction yield formed cyclic tetramers, namely, vanillin calix[4]arene for 32%, syringaldehyde calix[4]arene for 28%, and *p*-hydroxybenzaldehyde calix[4]arene for 37% [20].

Calixarene synthesis commonly needs a lot of solvents which results in a large amount of waste. Therefore, a synthesis approach by minimizing the use of solvents has been attempted and proved to be more effective. The solvent-free reactions that have been tried, in addition

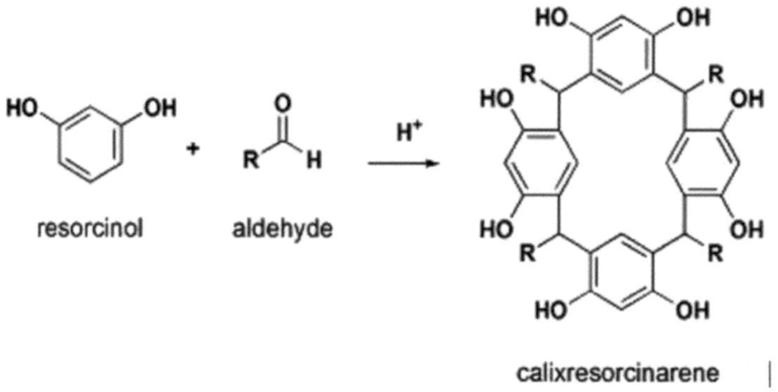


| Aldehyde | R | Calixresorcinarene |
|---|---|---|
| Vanillin (4-hydroxy-3-methoxyphenylbenzaldehyde) | 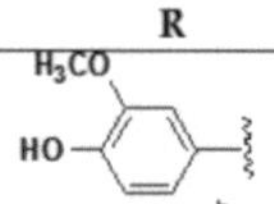 | C-4-hydroxy-3-methoxphenil calix[4]resorcinarene |
| Cinnamaldehyde (4-methoxyphenylbenzaldehyde) | 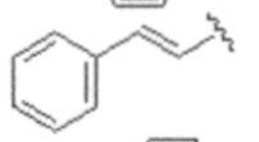 | C-2-phenylethenil calix[4]resorcinarene |
| Anisaldehyde | 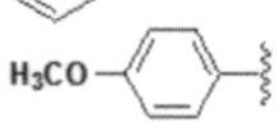 | C-4-methoxyphenyl calix[4]resorcinarene |

**Figure 12.** Synthesis reaction of calix[4]resorcinarene from resorcinol with various aldehydes.

to requiring a shorter amount of time, used less energy and could produce better and more effective and selective reactions than the conventional method [49]. Solvent-free reactions have also been used for calixresorcinarene synthesis, which is a condensation reaction between aldehyde and resorcinol. The method has also been widely used in several organic reactions and sometimes gives better results than the reaction method in solvents [50].

In calixresorcinarene synthesis, the use of solvent-free method was carried out by mixing aldehyde and resorcinol (1:1) with a certain amount of acid catalysts and then ground simultaneously. Although all reagents were in solid forms, when mixed they could form a thick liquid or paste reaction compound. Isolation of the reaction product can be carried out by washing with water to remove the acid catalyst and recrystallize with the right solvent. The method was proven to effectively yield calixresorcinarene with various aldehydes. In this method, no external heating is needed; only consistent scouring is needed. The scouring can be done in order to reduce particle sizes so that the compounding will be efficient. Because the reaction forms a thick liquid, not powder compound, then scouring at high intensity is not necessary [51].

Using the solvent-free method, the reaction occurs in mere minutes, much faster than the reaction time using the conventional method, which requires hours or even days. The percentage yield of the solvent-free method is often greater than that of the conventional synthesis method. The synthesis of calix[4]resorcinarenes commonly yields isomeric products, with two forms of isomer, namely, cis-cis-cis (rccc) and cis-trans-trans (rctt) with "crown" ($C_{4v}$), "chair" ($C_{2h}$), or "boat" ($C_{2v}$) conformation. Theoretically, without the effect of solvents, rccc isomers with "crown" ($C_{4v}$) conformation are more desirable than the rctt isomers with "chair" ($C_{2h}$) conformation [51]. Besides reduced reaction time, synthesis with this method uses a small amount of catalyst and does not produce solvent waste, hence meeting the principles of "green synthesis" [51, 52].

The condensation reaction between aldehyde and resorcinol (1:1) with *p*-toluenesulfonate acidic catalyst (5%) occurs for several seconds up to several minutes and yields derivatives of calix[4]resorcinarenes with rccc and rctt isomers. Four out of the six reactions were reported by Roberts et al. [51] to yield more rctt isomers with "chair" ($C_{2h}$) conformation than rccc isomers with "crown" ($C_{4v}$) conformation, with a ratio of 2:1. However, in one of the reactions, more rccc "crown" ($C_{4v}$) isomers were yielded for up to 95%. The distribution of isomer products is very possibly related to solubility that also plays a role in the reaction with the solvent-free method. In this reaction, increased reaction or heating time of the initial materials up to 85°C for 5 h did not cause an increase in the amount of rccc isomers yielded with "crown" ($C_{4v}$) conformation [51].

Employing the same method, Firdaus et al. [53] reacted vanillin with resorcinol (1:1) and yielded C-4-hydroxy-3-methoxyphenylcalix[4]resorcinarene for 52%, with a ratio of rccc isomers to rctt isomers of 1:5. However, when the aldehyde used was replaced with *p*-anisaldehyde, 63% of C-4-methoxyphenylcalix[4]resorcinarene was yielded with rccc isomers two times greater in number than rctt isomers. The distribution of the isomer products showed that each aldehyde has such a different characteristic that it can yield a different product composition. This difference may also be related to the solubility of the aldehyde used in the reaction compound. In addition, the reaction product demonstrated that rccc isomers with "crown" ($C_{4v}$) conformation are more desirable [53].

The solvent-free method is significantly faster than the conventional synthesis method. This can also be seen in the synthesis of calix[4]pyrogallolarene that is commonly done with condensation reaction of pyrogallol and aldehyde using alcohol solvent and acid catalyst and yields in rccc "cone" isomers. In the synthesis of calix[4]pyrogallolarene using the solvent-free method, an amount of aldehyde was added drop by drop to pyrogallol (1:1) with *p*-toluenesulfonate (3.7%) catalyst and scoured constantly. The reaction yielded vulnerable white solids in 2 min. The solids were easily scoured into soft yellow powder in 5 min. As was the case with the synthesis reaction of calix[4]resorcinarene derivatives, no heating was used in this synthesis [52].

The synthesis of C-(*p*-substituted phenyl) calix[4]resorcinarenes was carried out to see the interaction between host and guest using DMSO by mixing resorcinol with halogen-substituted benzaldehyde in para positions (1:1) under acidic conditions with an addition of 5% solid *p*-toluenesulfonate. The synthesis yielded a product at a percentage of 90–95% by scouring without using any solvents [54].

In the synthesis of calixresorcinarenes from resorcinol and aromatic aldehyde, Lewis acid is frequently used, such as $BF_3$-$OEt_2$, $AlCl_3$, and $SnCl_4$. For example, in the synthesis of resorcinarene *O*-acetates from 1,3-di(alkoxycarbonylmethoxy)benzenes with aromatic or aliphatic aldehydes (1:1) can be efficiently done at room temperature with the addition of $BF_3$-$OEt_2$ (boron trifluoride etherate) as a catalyst in order to produce high yields [55]. However, the amount of acid needed in this reaction is quite large (50–200% mole). Certainly, this is not good for the environment and safety, considering the sensitivity of Lewis acid to water that can yield side products in the forms of acidic oxides and metal oxides. Furthermore, the process is not efficient in terms of atom economy. The reaction also can yield two diastereoisomers. Lanthanide (III) triflates are selective acid catalysts that have been widely used for the formation of carbon-carbon and carbon-heteroatom bonds. Different from the common Lewis acid that is frequently used in stoichiometric numbers, lanthanide (III) triflates can be used in various reactions in catalytic numbers in the presence of solvents such as THF, DMSO, DMF, MeCN, and water. In addition, the stability and solubility of this catalyst in water make the compound easily separated and reused. One of the examples of lanthanide (III) triflate compounds that can be used as a catalyst is Ytterbium (III) triflate nonahydrate $\{[Yb(H_2O)_9](OTF)_3\}$. In the condensation reaction of resorcinol with various aldehydes (1:1), to produce derivatives of calix[n]resorcinarenes only needs 8% mole of Ytterbium (III) triflate nonahydrate as a catalyst. Although the amount of the catalyst is small, the reaction product yielded can reach up to 71–94% and this reaction is stereoselective. The reaction is done by refluxing for 48 h with ethanol as solvent [56].

Tungstate sulfuric acid (TSA) can also be used in the synthesis of calix[4]resorcinarene derivatives. In the reaction of an aldehyde with resorcinol at 120°C, the addition of TSA could produce around 78–95% in 25–80 min. The use of TSA as an acid catalyst is also more environmentally friendly because this acid can be recycled and reused for up to four reaction cycles with a decrease in reaction product yield around 3–11% [57].

# 3. Conclusion

Calixarenes and their derivatives can form a cyclic oligomer with a number of monomer units ranging from 4 to 20 and a basic structure consisting of phenolic units or phenol derivatives which are repetitive and interconnected by the Ar-C-Ar bridge to form a hollow cycle. Cyclic oligomer calixarene and its derivatives have been widely used in various fields, especially as host molecules in a guest-host system. Therefore, it takes a lot of effort to synthesize calixarenes and their derivatives. Various modifications of functional groups and synthesis methods were made in order to produce the expected configuration and conformation. These compounds have largely been synthesized by conventional methods by heating at 50–120°C for 1.5–45 h, either under acidic or basic conditions. The conventional method requires the use of various solvents in large amounts. The percentage yield of the reaction employing this method can range from 50 to 91%. The conventional synthesis method that requires heating at high temperatures, a fairly long reaction time, and large amounts of solvents has encouraged many researchers to develop more environmentally friendly synthesis methods, such as microwave-assisted synthesis and solvent-free methods.

Microwave-assisted synthesis of calixarenes and their derivatives requires 2–8 min with a percentage of reaction products of up to 99%. This method has been proven to accelerate reaction time without reducing the percentage of reaction products significantly. The synthesis of calixarenes and their derivatives can also be done with the solvent-free method. The method only relies on scouring process without heating. Nevertheless, the method requires a shorter reaction time with a higher percentage yield than that from the conventional synthesis method. Both microwave-assisted synthesis method and solvent-free method can reduce reaction time, energy use, solvent, and waste, making the synthesis of oligomer calixarenes and their derivatives more environmentally friendly.

# Author details

Ratnaningsih Eko Sardjono* and Rahmi Rachmawati

*Address all correspondence to: ratnaeko@upi.edu

Indonesia University of Education, Bandung, Indonesia

# References

[1] Gutsche CD. Calixarenes: An Introduction. 2nd ed. Cambridge: The Royal Society of Chemistry; 2008. 276 p. DOI: 10.1039/9781847558190

[2] Reinhoudt DN. Introduction and history. In: Neri P, Sessler JL, Wang MX, editors. Calixarenes and Beyond. Switzerland: Springer; 2016. pp. 1-11. DOI: 10.1007/978-3-319-31867-7

[3] Wu TT, Speas JR. Synthesis and characterization of a novel calix[4]arene tetramethyl tetraether. The Journal of Organic Chemistry. 1987;**52**(11):2330-2332. DOI: 10.1021/jo00387a050

[4] Iwanek W, Wzorek A. Introduction to the chirality of resorcinarenes. Mini-Reviews in Organic Chemistry. 2009;**6**(4):398-411. DOI: 10.2174/157019309789371604

[5] Jose P, Menon S. Lower-rim substituted calixarenes and their applications. Bioinorganic Chemistry and Applications. 2007;**2007**:1-16. DOI: 10.1155/2007/65815

[6] Sardjono RE, Dwiyanti G, Aisyah S, Khoerunnisa F. Synthesis of calix[4]resorsinarene from cinnamon oil and its application for solid phase extraction of heavy metals Hg(II) and Pb(II). (In Indonesia: Sintesis calix[4]resorsinarena dari minyak kayumanis dan penggunaannya untuk ekstraksi fasa padat logam berat Hg(II) dan Pb(II)). Jurnal Pengajaran MIPA. 2008;**12**(1):1-13. DOI: 10.18269/jpmipa.v12i1.318

[7] Timmerman P, Verboom W, Reinhoudt DN. Resorcinarenes. Tetrahedron. 1996;**52**(8):2663-2704. DOI: 10.1016/0040-4020(95)00984-1

[8] Kleinhans D. Studies in the Selective Synthesis of Bidentate Resorcinarene Ligands [thesis]. Stellenbosch: Stellenbosch University; 2010. 101 p. Available from: http://scholar.sun.ac.za/handle/10019.1/4194

[9] Sardjono RE. Sintesis kalisksarena dari minyak adas dan penggunaannya sebagai penjebak kation logam berat (In Indonesia: Synthesis of calixarene from dari minyak adas dan penggunaannya sebagai penjebak kation logam berat) [thesis]. Yogyakarta: Universitas Gadjah Mada; 1999.

[10] Yonetake K, Nakayama T, Ueda M. New liquid crystals based on calixarenes. Journal of Materials Chemistry. 2001;**11**(3):761-767. DOI: 10.1039/b005461p

[11] Yang F, Guo H, Vicens J. Mini-review: calixarene liquid crystals. Journal of Inclusion Phenomena and Macrocyclic Chemistry. 2014;**80**(3):177-186. DOI: 10.1007/s10847-014-0394-6

[12] Fang X, Guo H, Yang F, Wu Y. Novel gallic-calixarene liquid crystals: syntheses and conformation influences on mesomorphism. Tetrahedron Letters. 2015;**56**(44):6128-6131. DOI: 10.1016/j.tetlet.2015.09.093

[13] Tanwar S, Ho JA, Magi E. Green synthesis and characterization of novel gold nanocomposites for electrochemical sensing applications. Talanta. 2013;**117**:352-358. DOI: 10.1016/j.talanta.2013.09.011

[14] McMahon G, O'Malley S, Nolan K, Diamond D. Important calixarene derivatives - their synthesis and applications. Archive for Organic Chemistry. 2003;**2003**(7):23-31. DOI: 10.3998/ark.5550190.0004.704

[15] Tashakkorian H, Lakouraj MM, Rouhi M. p-Sulfonic acid calix[4]arene as an efficient catalyst for one-pot synthesis of pharmaceutically significant coumarin derivatives under solvent-free condition. International Journal of Medicinal Chemistry. 2015;**2015**:1-8. DOI: 10.1155/2015/738202

[16] Maksimov AL, Buchneva TS, Karakhanov EA. Supramolecular calixarene-based catalytic systems in the Wacker-oxidation of higher alkenes. Journal of Molecular Catalysis A: Chemical. 2004;**217**(1-2):59-67. DOI: 10.1016/j.molcata.2004.03.024

[17] Ding C, Qu K, Li Y, Hu K, Liu H, Ye B, Wu Y, Zhang S. Preparation and characterization of six calixarene bonded stationary phases for high performance liquid chromatography. Journal of Chromatography A. . 2007;**1170**(1-2):73-81. DOI: 10.1016/j.chroma.2007.09.036

[18] Yang F, Liu W, Xie J, Bai X, Guo H. Novel deep-cavity calix[4]arene derivatives with large s-triazine conjugate systems: synthesis and complexation for dyes. Journal of Inclusion Phenomena and Macrocyclic Chemistry. 2013;**76**(3):311-316. DOI: 10.1007/s10847-012-0200-2

[19] Namor AFD, Aparicio-Aragon W, Nwogu N, Gamouz AE, Piro OE, Castellano EE. Calixarene and resorcarene based receptors: from structural and thermodynamic studies to the synthesis of a new mercury (II) selective material. The Journal of Physical Chemistry B. 2011;**115**(21):6922-6934. DOI: 10.1021/jp110195f

[20] Nakajima L, Kobayashi T. Calixarene hosts synthesized from green chemicals of vanillin and syringaldehyde: fluorescence probes for Pb2+. Journal of Chemical Engineering of Japan. 2016;**49**(9):878-883. DOI: 10.1252/jcej.16we056

[21] Zadmard R, Taghvaei-Ganjali S, Gorji B, Schrader T. Calixarene dimers as host molecules for biologically important di- and oligophosphates. Chemistry - An Asian Journal. 2009;**4**(9):1458-1464. DOI: 10.1002/asia.200900085

[22] Shinkai S. Molecular recognition of calixarene-based host molecules. Journal of Inclusion Phenomena and Molecular Recognition in Chemistry. 1989;**7**(2):193-201. DOI: 10.1007/BF01060721

[23] Rebek J. Host-guest chemistry of calixarene capsules. Chemical Communications. 2000;(8):637-643. DOI: 10.1039/a910339m

[24] Kunsagi-Mate S, Nagy G, Kollar L. Host-guest interaction of calixarene molecules with neutral benzotrifluorides comparison of luminescence spectral data with results of model calculations relating to complex formation. Analytica Chimica Acta. 2001;**428**(2):301-307. DOI: 10.1016/S0003-2670(00)01237-X

[25] Bauer LJ, Gutsche CD. The formation of complexes of calixarenes with neutral organic molecules in solution. Journal of the American Chemical Society. 1985;**107**(21):6063-6069. DOI: 10.1021/ja00307a040

[26] Chaabane RB, Gamoudi M, Remaki B, Guillaud G, Beqqali OE. Investigation of the electrical properties of the metal-calixarene-semiconductor structures. Thin Solid Films. 1997;**296**(1-2):148-151. DOI: 10.1016/S0040-6090(96)09344-3

[27] Fareghi-Alamdari R, Golestanzadeh M, Zekri N. Solvent-free synthesis of trisphenols as starting precursors for the synthesis of calix[4]arenes using sulfonated multi-walled carbon nanotubes. New Journal of Chemistry. 2016;**40**(4):3400-3412. DOI: 10.1039/c5nj03136b

[28] Osipov M, Chu Q, Geib SJ, Curran DP, Weber SG. Synthesis of deep-cavity fluorous calix[4]arenes as molecular recognition scaffolds. Beilstein Journal of Organic Chemistry. 2008;**4**(36):6-11. DOI: 10.3762/bjoc.4.36

[29] Bell SEJ, Browne JK, McKee V, McKervey MA, Malone JF, O'Leary M, Walker A. Cation complexation by chemically modified calixarenes. 11. complexation and extraction of alkali cations by calix[5]-and-[6]arene ketones. Crystal and molecular structures of calix[5]arene ketones and Na + and Rb + complexes. The Journal of Organic Chemistry. 1998;**63**(3):489-501. DOI: 10.1021/jo9713114

[30] Hanna TA, Liu L, Angeles-Boza AM, Kou X, Gutsche CD, Ejsmont K, Watson WH, Zakharov LN, Incarvito CD, Rheingold AL. Synthesis, structures, and conformational characteristics of calixarene monoanions and dianions. Journal of the American Chemical Society. 2003;**125**(20):6228-6238. DOI: 10.1021/ja0289797

[31] Wang Z, editor. Comprehensive Organic Name Reactions and Reagents: Volume 3. New Jersey: Wiley; 2010. pp. 3183-3196. DOI: 10.1002/9780470638859

[32] Mendoza J, Nieto PM, Prados P, Sanchez C. A stepwise synthesis of functionalized calix[4]arenes and a calix[6]arene with alternate electron-withdrawing substituents. Tetrahedron. 1990;**46**(2):671-682. DOI: 10.1016/S0040-4020(01)85448-X

[33] Iqbal M, Stewart D. Calixarenes. 19. Syntheses procedures for p-tert-butylcalix[4]arene. The Journal of Organic Chemistry. 1986;**51**(5):742-745. DOI: 10.1021/jo00355a033

[34] Gutsche CD, Rogers JS, Stewart D, See KA. Calixarenes: paradoxes and paradigms in molecular baskets. Pure and Applied Chemistry. 1990;**62**(3):485-491. DOI: DOI: 10.1351/pac199062030485

[35] Prabawati SY, Jumina, Santosa SJ, Mustofa. Synthesis of polypropylcalix[6]arene from p-t-butylphenol as adsorbent for Cr(III) metal ion. Indonesian Journal of Chemistry. 2011;**11**(1):37-42. DOI: 10.22146/ijc.621

[36] Bourgeois J, Stoeckli-Evans H. Synthesis of new resorcinarenes under alkaline conditions. Helvetica Chimica Acta. 2005;**88**(10):2722-2730. DOI: 10.1002/hlca.200590211

[37] Utomo SB, Jumina, Siswanta D, Mustofa, Kumar N. Synthesis of thiomethylated calix[4]resorcinarene based on fennel oil via chloromethylation. Indonesian Journal of Chemistry. 2011;**11**(1):1-8. DOI: 10.22146/ijc.615

[38] Plachkova-Petrova D, Petrova P, Miloshev S, Novakov C. Optimization of reaction conditions for synthesis C-tetramethylcalix[4]resorcinarene. Bulgarian Chemical Communications. 2012;**44**(3):208-215

[39] Hogberg AGS. Two stereoisomeric macrocyclic resorcinol-acetaldehyde condensation products. The Journal of Organic Chemistry. 1980;**45**(22):4498-4500. DOI: 10.1021/jo01310a046

[40] Abis L, Dalcanale E, Vosel AD, Spera S. Structurally new macrocycles from the resorcinol-aldehyde condensation. Configurational and conformational analyses by means of dynamic NMR, NOE, and T1 experiments. The Journal of Organic Chemistry. 1988;**53**(23):5475-5479. DOI: 10.1021/jo00258a015

[41] Kappe CO. Controlled microwave heating in modern organic. Angewandte Chemie International Edition. 2004 ;**43**(46):6250-6284. DOI: 10.1002/anie.200400655

[42] Hoz A, Diaz-Ortiz A, Moreno A. Microwaves in organic synthesis: thermal and non-thermal microwave effects. Chemical Society Reviews. 2005;**34**(2):164-178. DOI: 10.1039/B411438H

[43] Gabriel C, Gabriel S, Grant EH, Halstead BSJ, Mingos DMP. Dielectric parameters relevant to microwave dielectric heating. Chemical Society Reviews. 1998;**27**(3):213-224. DOI: 10.1039/A827213Z

[44] Satyanarayana S, Kumar KP, Reddy PL, Narender R, Narasimhulu G, Reddy BVS. Microwave-assisted cyclocondensation: a rapid and solvent-free synthesis of 3-benzyl-2H-pyrido[1,2-a]pyrimidin-2-one derivatives. Tetrahedron Letters. 2013;**54**(36):4892-4895. DOI: 10.1016/j.tetlet.2013.06.138

[45] Baozhi L, Gengliang Y, Jinsong Z, Kefang D. Microwave-assisted synthesis of p-alkylcalix[n]arene catalyzed by KOH. E-Journal of Chemistry. 2005;**2**(1):70-74. DOI: 10.1155/2005/470510

[46] Hedidi M, Hamdi SM, Mazari T, Boutemeur B, Rabia C, Chemat F, Hamdi M. Microwave-assisted synthesis of calix[4]resorcinarenes. Tetrahedron. 2006;**62**(24):5652-5655. DOI: 10.1016/j.tet.2006.03.095

[47] Patel MH, Shrivastav PS. Microwave assisted synthesis of thiacalix[n]arenes. Journal of Inclusion Phenomena and Macrocyclic Chemistry. 2009;**63**(3):379-382. DOI: 10.1007/s10847-008-9520-7

[48] Sardjono RE, Kadarohman A, Mardhiyah A. Green synthesis of some calix[4]resorcinarene under microwave irradiation. Procedia Chemistry. 2012;**4**:224-231. DOI: 10.1016/j.proche.2012.06.031

[49] Makha M, Raston CL, Skelton BW, White AH. A more benign approach to the synthesis of calixarenes. Green Chemistry. 2004;**6**(3):158-160. DOI: 10.1039/B315204A

[50] Tanaka K, Toda F. Solvent-free organic synthesis. Chemical Reviews. 2000;**100**(3):1025-1074. DOI: 10.1021/cr940089p

[51] Roberts BA, Cave GWV, Raston CL, Scott JL. Solvent-free synthesis of calix[4]resorcinarenes. Green Chemistry. 2001;**3**(6):280-284. DOI: 10.1039/b104430n

[52] Antesberger J, Cave GWV, Ferrarelli MC, Heaven MW, Raston CL, Atwood JL. Solvent-free, direct synthesis of supramolecular nano-capsules. Chemical Communication. 2005;**7**(7):892-894. DOI: 10.1039/B412251H

[53] Firdaus M, Jumina, Anwar C. Green synthesis of C-4-Hydroxy-3-methoxyphenylcalix[4]resorcinarene, and C-4-Methoxyphenylcalix[4]resorcinarene. In: Proceeding of The International Seminar on Chemistry; 30-31 October 2008; Jatinangor. Bandung: Universitas Padjajaran; 2008. pp. 346-350

[54] Alshahateet SF, Al-Trawneha SA, Al-Zereinib WA, Al-Sarhana SS. Green synthesis, crystal structure and bioactivity of c-(*p*-substituted phenyl)calix[4]resorcinarenes-DMSO inclusion complexes. Jordan Journal of Chemistry. 2014;**9**(3):170-186

[55] Zhou R, Ren JC, Yan CG. Novel synthesis of resorcinarene O-acetates by BF3·OEt2-catalyzed cyclocondensation of 1,3-(dialkoxycarbonylmethoxy)benzenes with aldehydes. Journal of Inclusion Phenomena and Macrocyclic Chemistry. 2010;**67**(3):335-342. DOI: 10.1007/s10847-009-9713-8

[56] Barrett AGM, Braddock DC, Henschke JP, Walker ER. Ytterbium (III) triflate-catalysed preparation of calix[4]resorcinarenes: Lewis assisted Brønsted acidity. Journal of the Chemical Society, Perkin Transactions 1. 1999;**1999**(8):873-878. DOI: 10.1039/A809919G

[57] Karami B, Khodabakhshi S, Safikhani N, Arami A. A green and highly efficient solvent-free synthesis of novel calicx[4]resorcinarene derivatives using tungstate sulfuric acid. Bulletin of the Korean Chemical Society. 2012;**33**(1):123-127. DOI: 10.5012/bkcs.2012.33.1.123

# Environment-Friendly Approach in the Synthesis of Metal/Polymeric Nanocomposite Particles and Their Catalytic Activities on the Reduction of *p*-Nitrophenol to *p*-Aminophenol

Noel Peter Bengzon Tan and Cheng Hao Lee

Additional information is available at the end of the chapter

http://dx.doi.org/10.5772/intechopen.68388

**Abstract**

In this chapter, an environment-friendly approach in synthesizing Au and Au@Ag metal nanoparticles using a microgel is demonstrated. Poly(*N*-isopropyl acrylamide)/polyethyleneimine microgel was used as a multifunctional template to reduce metal ions to metal nanoparticles, stabilize and immobilize these metal nanoparticles, and regulate their accessibility within the template. Such multifunctional roles of microgel template were possible due to their unique properties (i.e., amino groups reducing capability, electrostatic and steric stabilizing properties, and swelling/deswelling properties). Characterizations of these metal/polymeric composite particles were also performed, such as scanning electron microscope (SEM), transmission electron microscope (TEM), Zeta-potential, UV-vis spectroscopy, X-ray Diffraction (XRD), and X-ray photoelectron spectroscopy (XPS). To test the catalytic activities of both gold and gold@silver nanoparticles in the microgel template, a model reaction (i.e., reduction of *p*-nitrophenol to *p*-aminophenol) was performed. Results showed that bimetallic gold@silver gave 10 times higher catalytic activity compared to monometallic gold nanoparticles. With the simple one-step synthesis, a highly scalable process is possible.

**Keywords:** green synthesis, gold nanoparticles, Au@Ag bimetallic nanoparticles, core-shell particles, smart microgel particles, smart materials

## 1. Introduction to metal/polymeric nanocomposite particles

Metal/polymeric nanocomposite particles are a combination of both metal particles and polymers in nanoscale. They come in different terms and play of words. But simply they are colloidal polymers with metal nanoparticles. Metal nanoparticles that can be incorporated into different colloidal polymeric systems are magnetic, semiconductor, and noble metals. On the other hand, colloidal polymers act as carriers of these metal nanoparticles. They are mostly referred to as polymer templates. These templates can either be soluble (i.e., colloidally soluble) or insoluble (i.e., solid or heterogeneous) polymers. **Figure 1** displays the different conformations of metal nanoparticles with polymeric templates. For example, metal nanoparticles are seen as core (**Figure 1a** and **b** [1–2]) or part of the polymeric template shell (**Figure 1c** [3]) or attached to both the core and shell of the composite (**Figure 1d** [4]).

Applications of metal/polymeric nanocomposites vary from the fields of chemistry (e.g., catalysis, sensors, and polymers), physics (e.g., optics and electronics), biology (e.g., detection and control of microorganism), and nanomedicine (e.g., drug development and immunoassay).

There are two general approaches in synthesizing metal nanoparticles: top-down and bottom-up. Top-down methods comprise physical methods such as lithography and etching of bulk metals to nanoscopic scale. Bottom-up approaches are more common these days than the top-down. The bottom-up approach also has an advantage of generating uniform nanoparticles with controlled size and shape.

Bottom-up approaches or commonly referred to as wet chemical methods were pioneered for more than a century ago [5]. In particular, Michael Faraday's method used metal-salt solution mixed with reducing agents (e.g., hydrogen, alcohol, hydrazine, or borohydride)

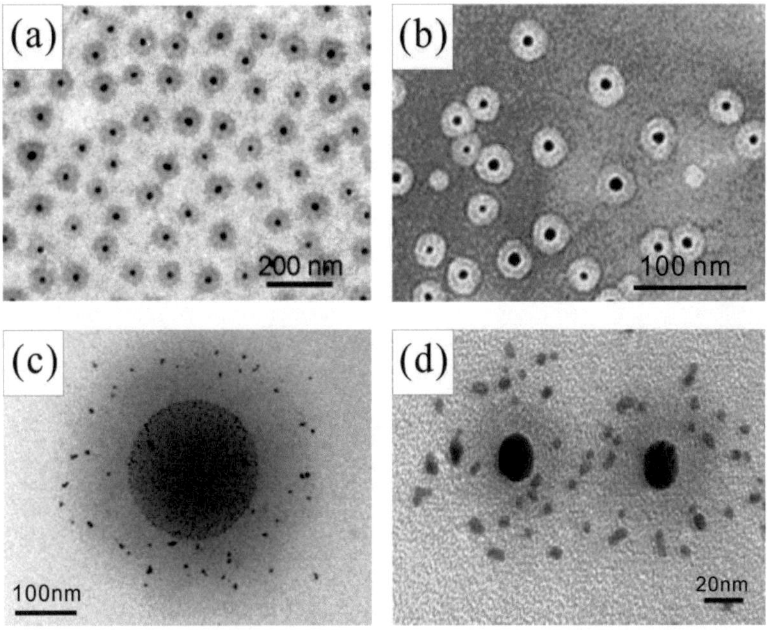


**Figure 1.** TEM images of metal nanoparticles (dark dots) encapsulated within its polymeric templates: (a) PNIPAm-*b*-PMOEGMA/Au [Taken from Ref. [1]], (b) PS-*co*-PGA/Au [Taken from Ref. [2]], (c) PS/poly (aminoethylmethacrylate HCl)/gold particles PNIPAm-*co*-GMA/Au [Taken from Ref. [3]], and (d) PNIPAm-*b*-PMOEGMA/ Au with Ag [Taken from Ref. [4]].

http://dx.doi.org/10.5772/intechopen.68388

and later, stabilizing agents (e.g., ligands, polymers, or surfactants). Turkevich et al. [6] and Brust-Schiffrin et al. [7] were able to use this similar approach by synthesizing gold nanoparticles. Their synthetic route involved the reaction of a chloroauric acid with sodium citrate solution at boiling temperature ($HAuCl_4 + Na_3C_6H_5O_7 = Au^0$). Later on, Frens [8] was able to control the size formation of gold nanoparticles by varying the reducing agent to gold-salt ratio during the reduction process. Furthermore, Yonezawa and Kunitake [9] used sodium 3-mercaptopropionate to the prestabilized citrate gold nanoparticles. The Brust-Schiffrin's method involves the reduction of gold-salt solution using a thiol-based organic solvent in a two-phase system. The organic layer is separated, evaporated, and mixed with ethanol to get rid of excess thiol. The crude product is further dissolved in toluene and precipitated in ethanol. A modified Brust-Schiffrin process was carried out by Murray et al. [10] or commonly called as "place exchange" process. This process used various functionalities, such as bromine, cyanide, ferrocenyl, alcohol, formaldehyde, and anthraquinone, in replacement of a simple alkane group. Sulfur ligands such as xanthates [11], disulfides [12], di and trithiols [13], and resorcinarene tetrathiols [14] have also been utilized for gold nanoparticles (AuNPs) syntheses. Biphasic methods of AuNP synthesis can also use similar ligands such as phosphine [15], amine [16], carboxylate [17], isocyanides [18], citrate with acetone [19], and iodine [20]. Structures of these ligands are shown in **Figure 2**.

In general, there are three classifications of approaches to prepare AuNPs: (1) nonchemical methods such as electrochemical [21] and thermal decomposition of a metal-salt solution [22], photochemical [23], sonochemical [24], laser ablation synthesis [25], and microwave-assisted technique [26]; (2) biological sources such as the use of plant extracts and microorganism-assisted formation of metal nanoparticles. Bio-reduction of metal ions involves both intracellular and extracellular precipitations of metal nanoparticles within the microorganism (**Figure 3**) [27]. Biomolecules such as proteins are mainly responsible for the synthesis of gold nanoparticles while enzymes produced in the outer layer membrane of the microorganism are responsible for the reduction of gold ions. The biological pathways for metal nanoparticles synthesis

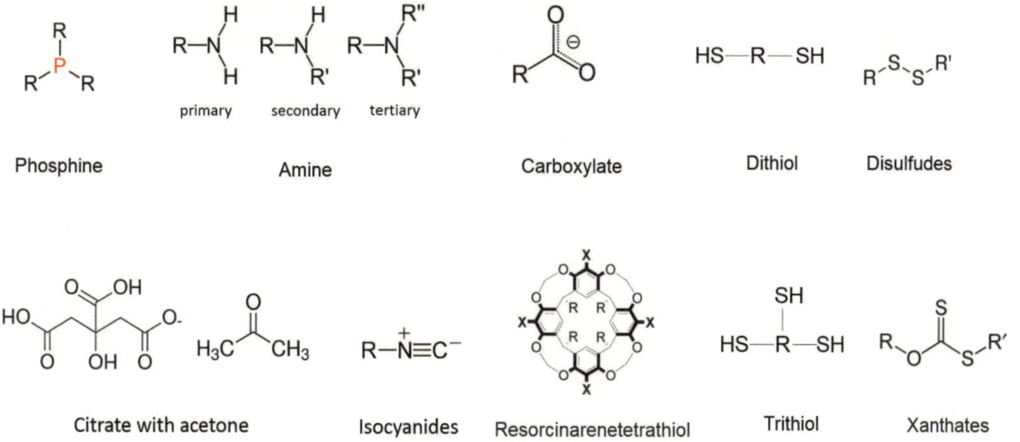


**Figure 2.** Different ligand molecular structures that can be used for gold nanoparticle synthesis.

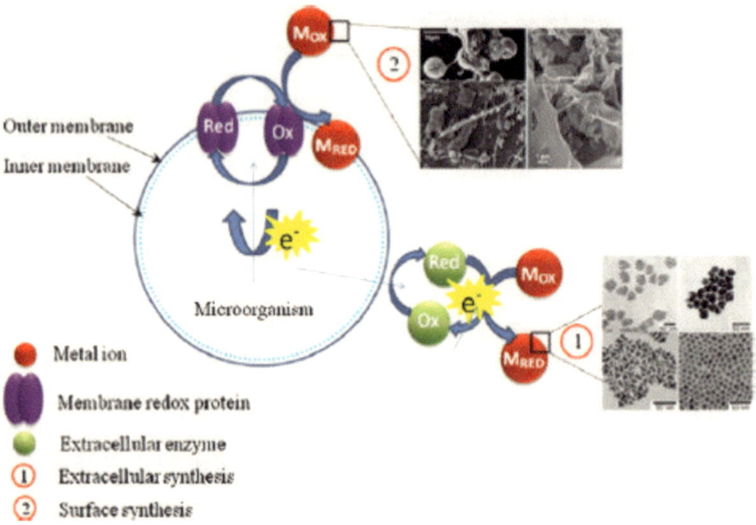


**Figure 3.** Biosynthetic mechanism of metal nanoparticles using microorganism [Taken from Ref. [27]].

can be carried out by a microorganism (e.g., bacteria [28], yeast (*P. jadini*), and fungal (*V. luteo-album*) cultures [29]), plants [30–32] and plant extracts [33]. In recent years, development of plant extract-based synthesis of metal nanoparticles has been investigated. Using plant-based synthesis results into more stable and faster rate of synthesis compared in the case of micro-organism [34]. (3) Use of a polymer as a template for metal nanoparticles (MNPs) generation is commonly called polymer-mediated synthesis. This emerging type of approach was conceived to solve the issue on MNP aggregation.

### 1.1. Polymer-mediated synthesis of metal nanoparticles

Polymers that have both reducing and stabilizing properties have been developed to synthesize metal nanoparticles. Such dual properties give pure and homogenous products. The main feature of this approach lies on its low cost, high efficiency, and environmentally benign nature. Several existing polymers, which display these dual properties (e.g., reducing and stabilizing metal nanoparticles), have already been used in the synthesis of MNPs such as poly(N-vinyl-2-pyrrolidone) (PVP) [35], poly(allylamine) (PAAm) [36], poly(o-phenylenediamine) (PoPD) [37], polyethyleneimine (PEI) [38], and poly(4-styrenesulfonic acid-co-maleic acid) (PSSMA) [39]. Mechanisms have been studied in the reducing capacity of the PVP. These include a free radical mechanism, oxidation of the hydroxyl end groups [40] and the C=O double bond [41]. Other factors include an abundance of amino groups in the PAAm and PEI molecules that drive the reduction of gold ions into metal nanoparticles and strong bonding between the electrons donor, π orbitals donor, and the lone pair orbitals of amine groups of PoPD with the electron-deficient orbitals of gold nanoclusters providing efficient stabilizing effect. Due to the high impact polymer-assisted approach on the synthesis of metal nanoparticles, several studies came up with some concluded advantages.

(1) Only a small concentration of polymer is used. (2) The functional groups in the polymer can serve for dual properties. (3) Polymer template itself can control the size and morphology of MNPs and its resultant composite.

### 1.2. Core-shell particles (CSP)

With the promising potential of the polymer-assisted approach on the synthesis of metal nanoparticles, the authors make use of polymeric particles. Here, it is referred as core-shell particles (CSP). Some of the commonly used polymers as templates and nanoreactors for metal nanoparticle formation are poly(glycidyl methacrylate-co-N-isopropylacrylamide) [(poly(GMA-co-NIPAM))] [42], poly(N-isopropylacrylamide)-co-poly(acrylic acid) (PNIPAM-co-PAA) [43], glycidyl methacrylate (GMA) and *N*- isopropyl acrylamide (NIPAM) [44], polystyrene (PS) core and a polyaniline (PANI) [45], (poly(N-isopropylacrylamide-acrylic acid) P(NIPAM-AA) [46], long cationic polyelectrolyte chains of poly(2-aminoethyl methacrylate hydrochloride) (PAEMH)) [47], and poly(ionic liquid) (PIL) [48].

Over the past decade, a metal-salt reduction process is the most common method for generating metal nanoparticles. This type of reaction has shown reliability and uniformity of metal nanoparticles produced. However, environmental concerns are not well addressed or worst not met. For example, the use of different forms of energy (e.g., photoirradiation, ultrasound irradiation, and high temperature boiling process) in both electrochemical and thermal decomposition methods is far way exploited [49] in addition to long and tedious synthetic procedures. And worst, giving low yields [50] with a high polydispersity of metal nanoparticles. Such high polydispersity is mostly observed in a reverse microemulsion of metal nanocomposite [51]. For metal nanoparticles bound ligands, the consequence of the difficulty in dispersing in water hinders the surface modification and functionalization for further applications [7]. As a result of this water incompatibility, metal nanoparticle properties are altered [52]. Also, some reducing agents such as sodium borohydride and hydrazine are considered toxic chemicals and not tolerable for future commercial scale-up [53]. Else, defective products or impurities may arise from excess reducing agents [54]. As a result, impurities left behind may eventually affect the composite material's functionality and its potential applications.

With the existing and emerging technologies in the synthesis of metal/polymeric nanocomposites, there is still a great challenge to the concept of Green Chemistry. This concept aims at the development of methods for the synthesis of metal/polymeric in this case, with the least impact on humans and environment as a whole. The challenges in creating novel metal/polymer nanocomposites are: (1) to create a unique template that is an all-in-one platform that can reduce metal ions to nanoparticles, immobilize the resultant nanoparticles, and stabilize the composite particle; (2) to regulate the accessibility of the metal nanoparticles through controlling external stimuli such as pH, temperature, and electrolyte; (3) to immobilize other organic and biological molecules for protection and deliveries; (4) to easily be purified and recovered; (5) to efficiently scale up process for commercialization.

# 2. Core-shell microgel template and metal/polymeric nanocomposite synthesis

A novel approach was developed with a simple yet versatile synthesis of a variety of amphiphilic core-shell particles [55]. This approach enables to synthesize a broad range of core-shell particles with different chemical structure, composition, size, and functionality. The process uses aqueous-based Chemistry, which is environmentally benign, and the particles are easy to synthesize in high solids content (up to 30%) in the absence of surfactant. A novel feature of this synthetic approach is that it combines graft copolymerization, *in situ* self-assembly of the resulting amphiphilic graft copolymers and emulsion polymerization in a one-step synthesis. In this chapter, core-shell microgels were used in the synthesis of mono (Au) and bimetallic (Au@Ag) nanoparticles. Briefly, the mechanism involved in the core-shell microgel synthesis combines graft copolymerization of vinyl monomer from a water-soluble polymer containing an amino group and self-assembly of the resulting particle. Graft polymerization of vinyl monomer in water happens when amino radicals are formed. The electron transfer and loss of proton form amino radicals during the interaction of alkyl hydroperoxide (ROOH) with the amino group of the polymer backbone (i.e., PEI is mostly used). Alkoxyl radicals (RO) are inevitably produced during this interaction. The resulting amphiphilic macroradicals undergo self-assembly forming micelle-like microdomains, where they become loci for the further polymerization of the monomers. The generated RO·radical on the other hand initiates homopolymerization of the vinyl monomer or creates radicals for further graft polymerization. This process results in well-defined core-shell particle structure with a hydrophilic shell and a hydrophobic core.

## 2.1. Synthesis of AuNPs in PNIPAm/PEI microgel template

The preparation of the Au nanocomposite (Au/(PNIPAm/PEI)) particles was carried out based on a previous method [56] developed by Tan et al., performed via the addition of hydrogen tetrachloroaurate(III) trihydrate ($HAuCl_4{\cdot}3H_2O$) solution into the as-prepared PNIPAm/PEI. The mixture was continuously stirred and carried out at different temperatures and pHs for 2 hours and heated at 60°C for about an hour. The resulting gold loaded microgels were then purified by centrifugation.

Gold nanoparticle formation in a microgel template is shown in a schematic diagram (**Figure 4**). Such formation from its ionic form is considered to be thermodynamically stable, which does not need any activation energy to form gold nanoparticles even at room temperature. Two successive reactions occur to complete the gold generation. First is the interaction between the negatively charged gold chloride ions ($AuCl_4^-$) and the cationic microgels. Once the gold ions are attracted into the microgel, a subsequent redox reaction occurs between the gold ions and available amine groups in the microgel template. As a result, gold ions are reduced while the amine groups are oxidized. Amine oxidation allows transfer of electrons from the amine to the gold ions, thus, generating zero-state AuNPs. Such reaction was reported by Lala et al. [57], wherein they have proposed that the $AuCl_4^-$ ions are electrostatically bound to

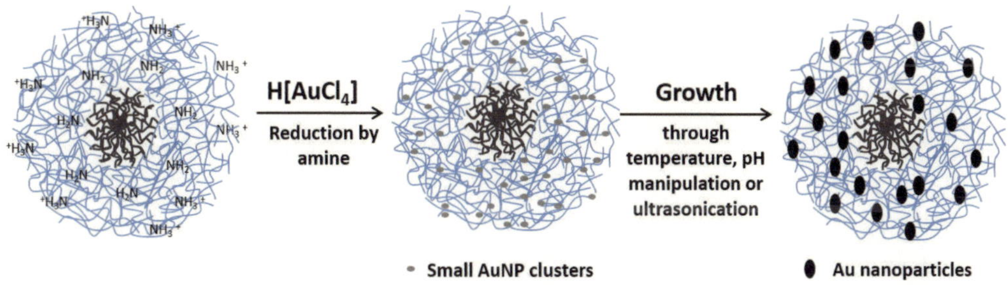


**Figure 4.** Schematic diagram on the synthesis of gold nanoparticles (AuNPs) using PNIPAm/PEI microgel as a template.

the protonated amine group and simultaneously reduced by the unprotonated amine group. AuNPs clusters were allowed to grow by further heating or manipulating their temperature or pH conditions.

### 2.2. Synthesis of Au@Ag/core-shell PNIPAm/PEI microgel composite particles

Au@Ag bimetallic nanoparticle synthesis was carried out through a progressive reduction of Au and Ag metal ions as performed previously by Tan et al. [58]. Gold metal ions were first reduced to the shell component of the microgel. These gold metal nanoparticles were then used as a seed for the successive reduction of the silver ions to silver nanoparticles. Appropriate molar ratios of $Au^{3+}$ and $Ag^{1+}$ ions were used and mixed for 30 minutes to reduce the silver ions to metal nanoparticles further, followed by heating at 60°C for 30 minutes.

Generating bimetallic nanoparticles in a microgel template is shown in a schematic diagram (**Figure 5a**). After the synthesis of microgel template through graft copolymerization, gold clusters were first generated on the shell layers of the templates. Such generation is possible due to hyperbranched PEI in the shell region, which contains amine groups that are known to have reducing ability to generate metal nanoparticles [59]. And through the chelating properties of the same amino groups, PEI can also complex with metal ions and metal nanoparticles [60]. The preformed gold nanoparticles acted as seeds or nucleation sites for further bimetallic nanocrystals formation. Such formation of Au seeds occurred after 30–40 minutes of reaction at room temperature which was evident by the change of the solution color from turbid white to light pink. The transition of the solution color also signifies the change in ionization potential and electron affinity values of Au atoms. Au atoms' ionization potential becomes higher than those of Ag atoms. Such shift results to a larger electronegativity value for Au, wherein significant charge transfer may occur from silver to gold atoms [61]. Simultaneously, silver metal ions ($Ag^+$) were reduced to silver nanoparticles through under-potential deposition mechanism [62], or noble metal induced reduction (NMIR) method [63]. Further, illustration of this mechanism is displayed in **Figure 5b**. It is shown that gold nanoparticles used as a seed for further reduction of silver ions to silver nanoparticles. The AuNP with a bigger size attracts the silver ions resulting to a bimetallic alloy nanoparticle. Further heating was necessary to improve the crystallinity of the bimetallic nanoparticles. Consequently, heating of these composite particles removes partially the template resulting in the naked exposure bimetallic nanoparticles.

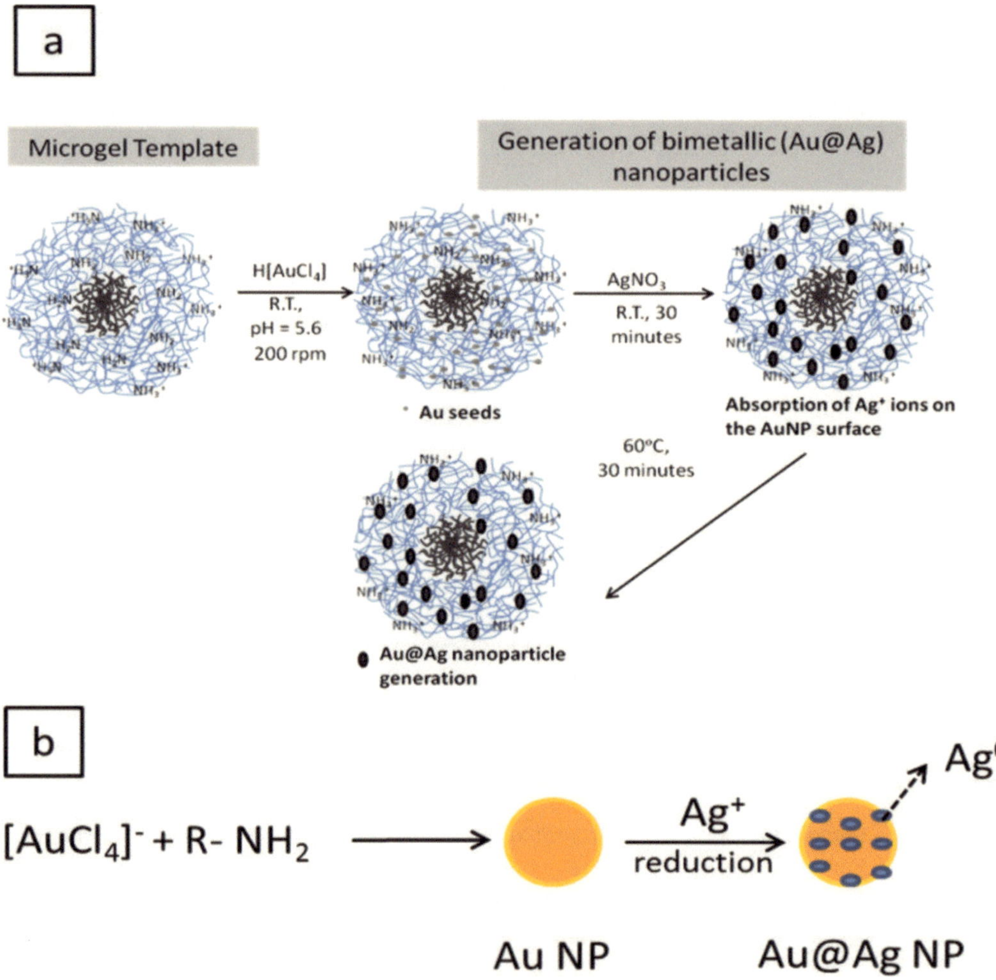


**Figure 5.** (a) Schematic diagram on the synthesis of bimetallic nanoparticles from Au/PNIPAm/PEI composite particles, (b) mechanism on the formation of Au@Ag nanoparticles from Au/PNIPAm/PEI nanocomposites.

## 3. Multifunctional roles of PNIPAm/PEI microgel

### 3.1. Microgel as a nanoreactor

Transmission electron microscope (TEM) images of both the empty PNIPAm/PEI microgel template and the gold nanoparticle-filled composite particles are shown in **Figure 4**. Herein, the empty PNIPAm/PEI microgel particles show a core-shell structure (**Figure 6a**), while AuNP-filled microgel template (**Figure 6b**) shows darks spots around its perimeter. The gold nanoparticles within the microgel template look like clusters of small gold nanoparticles. When heated to 60°C for an hour, the gold nanoparticles further crystallized and became clearer. The size of the gold nanoparticles was roughly estimated at an average of 17.60 ± 2.34 nm with a narrow size distribution.

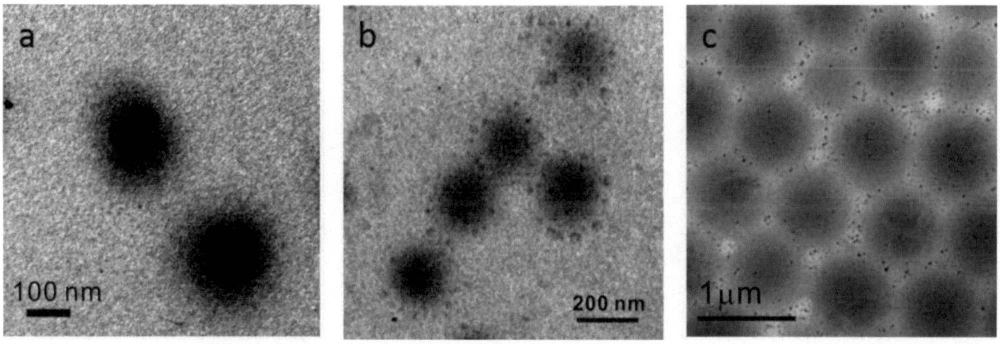


**Figure 6.** TEM images of the (a) PNIPAm/PEI microgel template, (b) Au/(PNIPAm/PEI) composite particles synthesized at 25°C and pH 7.30, and (c) heated Au/PNIPAm/PEI composite particles.

The kinetics of the formation of gold nanoparticles was monitored through the UV-vis absorbance at 525 nm wavelength with time. Such adsorption at 525 nm wavelength is one characteristic of AuNPs. In **Figure 7**, the increase of the absorbance was fast in the first 30 minutes of reaction and became slower after that, until the third hour of reaction. The reaction started to cease after 3 hours, and no significant change of absorbance was further observed. This data concludes that both electrostatic interaction and reduction of gold ions to nanoparticles simultaneously occurred at a fast rate. This graph further proves that there is a rapid nucleation during the initial stage of gold-ion reduction, resulting in numerous Au clusters. It was further concluded that the reduction of gold ions to AuNPs using microgel was 625 times faster than the naked hyperbranched PEI (linear curve with hollow points).

### 3.2. Microgel as a stabilizer of AuNPs in composite particles

There are two kinds of stabilization that holds both the AuNPs and the composite material in suspension. Such stabilization is due to the microgel template's property to provide electrostatic interaction between composite particles and steric effect of the PEI shell.

Electrostatic interaction between the composite particles and the gold nanoparticles within the template is the primary contributor to its stabilization. When Au nanoparticles are formed and immobilized in the particle template, the overall size of the composite particle becomes smaller than the pure template itself. This shrinkage is due to the formation of gold/amine complexes resulting in the contraction of the PEI shell. Such contraction of PEI shell reduces metal nanoparticles leaking from its template or its individual network-cage-like structure. Consequently, continuous leaking of naked AuNPs will form aggregates within the template. On the other hand, the same repulsion force acts between composite particles. Such force prevents them from getting attracted to each other preventing them from forming precipitates eventually.

Steric contribution to the stabilization of the AuNPs comes from the hyperbranched structure of the PEI-shell component in the microgel template. This type of stabilization is known in a lot of amphiphilic graft copolymers [64]. Such property of amphiphilic copolymers is due to the hydrophobic-hydrophilic interaction of the copolymers involved. This interaction is very

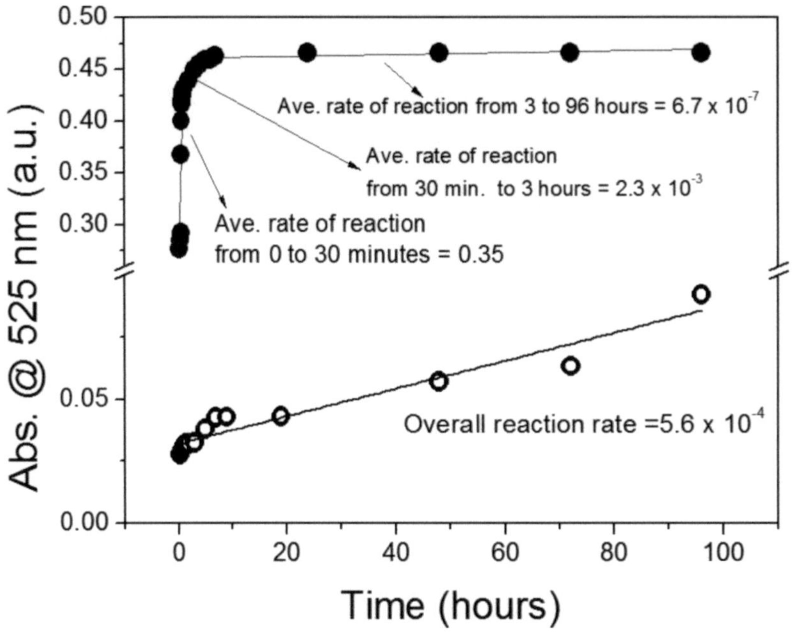


**Figure 7.** Time courses of the absorbance monitored at 525 nm during the formation of gold nanoparticles in the presence of PNIPAm/PEI microgel particles (25°C, at pH 5.6).

significant on the stability of the microgel template itself and in the formation of Au/microgel composite particles. When this interaction happens in the microgel template, the PEI shell anchors in the gold nanoparticles, while the PNIPAm core is kept together away from the shell. With such action, both the shape and stability of the composite particles are achieved.

### 3.3. Microgel as AuNP immobilizer

The generated gold nanoparticles using microgel template were immobilized through the template PEI shell's properties. Primarily, the weak bonding between the amino group and the gold nanoparticles is the primary source of immobilization [65, 66]. Such immobilization strongly supported by the hyperbranched nature of the PEI [67], which helps to shield AuNPs into a network-cage like structure. Such construction provides bulkiness and prevents the AuNPs from aggregating with neighboring AuNPs or composite particles. Furthermore, PEI-shell structure can also link the gold nanoparticles intact [68] within its boundary template.

There are five pieces of evidence to demonstrate the microgel acting as an immobilizer of gold nanoparticles: (1) In **Figure 6b**, AuNPs are seen as fuzzy gray dots embedded within the circumference of the microgel, attached in the shell region. (2) There was a decrease in the size of the pure microgel template when loaded with AuNPs. The decrease in size was due to the encapsulation of the gold metal ions attracted to the template. Absorption of the gold metal ions leads to the shrinking of the composite material. (3) There was a decrease in the zeta-potential from 30 to 15 mV from a pure microgel to Au-loaded template, respectively. Such decrease of the zeta-potential is attributed to the partial consumption of the cationic ammonium ions during the gold-ion adsorption stage. (4) X-ray photoelectron spectroscopy (XPS) (**Figure 14**) result further shows proof of the immobilization of AuNP in microgel template. This result verifies the location of AuNPs which are found within 2—10 nm deep from the surface of the microgel template. (5) The ligand role of the PEI shell (i.e., complexation of the water-soluble PEI with metal ions) plays a significant part of the immobilization of AuNPs. This ligand role property results in some advantages of the composite material such as water solubility, high capacity for metal uptake, easy separation of polymer complexes, high flexibility of the molecular conformation, and good chemical and physical stability [69–71].

### 3.4. Microgel as a smart controller of AuNP accessibility

One of the best features of PNIPAm/PEI microgel template is its ability to regulate its size. Such ability is useful in the accessibility of the gold nanoparticles generated within the microgel template. This ability of the microgel comes from the stimuli-responsive nature of the PNIPAm or some refer them to smart materials. In the case of PNIPAm/PEI microgel, such sensitivity is based on both sensitive pH and temperature. The core part of the microgel, PNIPAm is temperature sensitive, while the PEI shell is pH sensitive. The response of this soft template to temperature or pH affects its conformational structure. The changes in the structure of the template result in the controlled accessibility of AuNPs as demonstrated in **Figure 8**. Herein, the microgel template loaded with AuNPs is in different sizes under different pH or temperatures. At low pH, the template gets protonated and swells. Such swelling exposes the encapsulated AuNPs. However, when pH increases, the microgel becomes deprotonated and deswelling of the template occurs. By this action of the microgel template, AuNPs embedded within are trapped. The same action also controls that degree of plasmon coupling of AuNPs. Such coupling property originates from the dipole interaction among gold nanoparticles, which allows the control of the interparticle distance between gold nanoparticles [72].

On the other hand, when the temperature of the microgel system reaches beyond the lower critical solution temperature (LCST) point of the core part, PNIPAm (i.e., 32°C), the entire template shrinks. Such shrinking leads to the trapping of AuNPs within the template. But when the temperature goes below the LCST of PNIPAm, the template is more open and loose than the original condition. This looseness results in easy accessibility of the AuNPs within the template.

Microgel particles were subjected to different temperature conditions at 29°C, 45°C, and back to 29°C in aqueous solution to demonstrate the smart properties of the template. Their corresponding structural changes of the microgel particles under different temperatures were captured with AFM analysis. Original microgel template at 29°C in a fluid mode is shown in **Figure 8a** with sizes ranging from 100 to 150 nm with quasi-spherical morphologies. When the temperature was raised to 45°C (**Figure 8b**), the templates decreased in size showing porous surfaces. Such phenomenon is attributed to the shrinking of the templates as it goes beyond its volume phase transition temperature (VPTT). However, when restored to 29°C (**Figure 8c**), the smooth morphology and size of the templates were restored. Such restoration demonstrates that the conformational changes of the template triggered by the response to temperature are reversible.

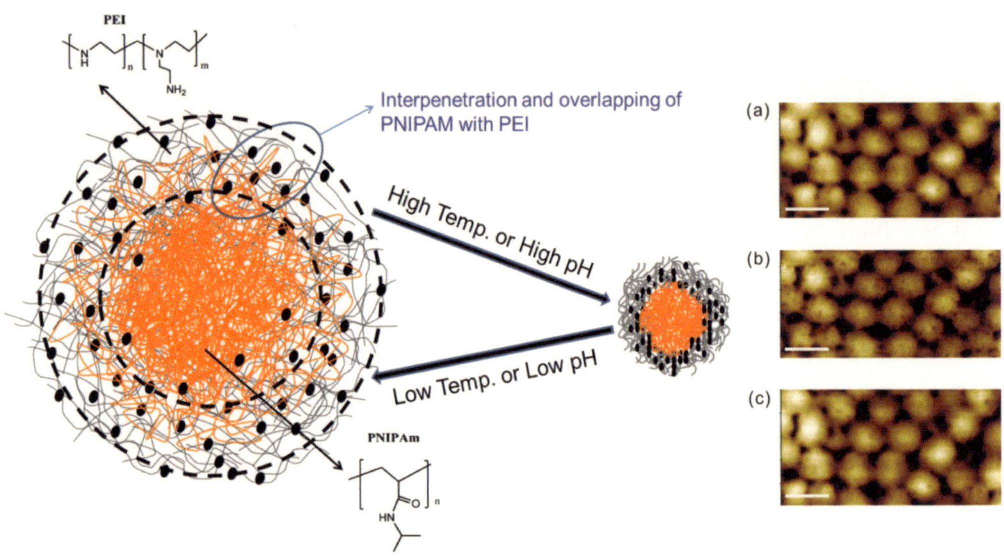


**Figure 8.** Left side: conformational changes of microgel template from stimuli response to pH solution and temperature. Right side: AFM micrographs of PNIPAM/PEI microgel particles measured in a fluid mode at different temperatures: (a) 29°C; (b) 45°C; and (c) Cooled from 45 to 29°C. Scale bar: 200 nm.

## 4. Measurements and characterization

### 4.1. Particle size and surface charge

Dynamic Light spectrophotometer measured the sizes of both pure and gold-loaded microgels. Synthesized PNIPAm/PEI microgels have an average hydrodynamic diameter of 402 nm while the gold-microgel composite particles were measured at 298 nm as shown in **Figure 9a**. The polydispersity indices on both unloaded and gold-loaded particles were 0.050 and 0.055, respectively. As anticipated, the particle size of the gold-loaded microgel is smaller than the pure microgel. This decline in size is due to the incorporation of the counterions into the microgel template during the metal ion absorption and reduction stages. Furthermore, when AuNPs are formed, the microgel network immobilizes *in-situ* generated AuNPs by capturing them on its network-like structure, providing a steric effect on the metal nanoparticles.

The gold-loaded microgel particles were further characterized based on its surface charge expressed in zeta-potential. Gold-loaded composite particles have an average zeta-potential of 15 mV at pH 7.00 in an aqueous medium. At this state, composite particles were stable with no aggregation or precipitation occurred. However, zeta-potential can be affected by the pH solution in a colloidal system. To demonstrate this effect, **Figure 9b** demonstrates the change of the surface charge as a function of pH. In the same figure, gold-loaded microgels can be grouped into a three-phase behavior regarding zeta-potential versus pH solution. The first phase shows a constant zeta-potential behavior at a pH range of 2–6.5. The second phase is between pH 6.5 and 9.0, which shows a noticeable decrease of zeta-potential. The third

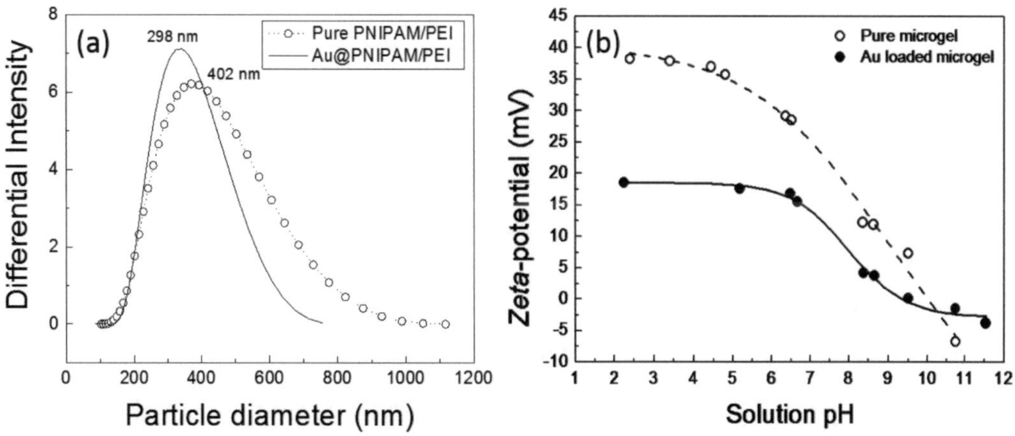


**Figure 9.** (a) Particle size and size distribution of pure microgel template and Au/(PNIPAm/PEI) composite particles synthesized at optimum conditions of 25°C and pH 7.30. (b) Zeta-potential profile of gold-loaded microgel (solid points) and pure microgel template (hollow points) in different pH solution.

phase between pH 9.0 and 11.5 gives a slight change of zeta-potential values. The constant zeta-potential in the first phase is attributed to the saturation of microgel template with AuNP at this certain range of pH. However, increasing the pH affects the composite material and decreases its surface charge surpassing the isoelectric point (i.e., pH 9.2). Further increase of pH at this stage may supersaturate the microgel template and then again give a very minimal or no effect on its zeta-potential.

To demonstrate the effect of temperature on its surface charge, **Figure 10** shows that varying solution temperatures from 25 to 40°C strongly affect the zeta-potential of both the pure and gold-loaded microgel particles. An abrupt change of surface charge in the temperature range between 29 and 34°C is obvious. This region crosses the VPTT region of the microgels. However, prior and after this temperature range, the zeta-potential was more or less constant. Such behavior is attributed to the increase in the surface charge density of the composite particles with the decrease in size. Smaller particles result in higher surface charge density, resulting in the shrinking of the composite particles, as also observed in the work of Ou et al. [73].

### 4.2. Scanning electron microscope (SEM) and transmission electron microscope (TEM) images

SEM and TEM images of the AuNP/(PNIPAm/PEI) composite particles are both shown in **Figure 11**. **Figure 11a** shows uniform spherical morphologies of the composite particles. Such morphologies are identical to that of the original microgel template (**Figure 11a** inset). However, partial agglomeration of the particles is also observed which may have occurred during the drying of the SEM sample treatment. **Figure 11b** shows the TEM image of the gold-loaded microgel which shows clearly the location of the AuNPs within the microgel template. Specifically, AuNPs reside around the circumference of the microgel attached in the shell region. Apparently, these images also show the effectiveness of the immobilization of the gold nanoparticles within the microgel network.

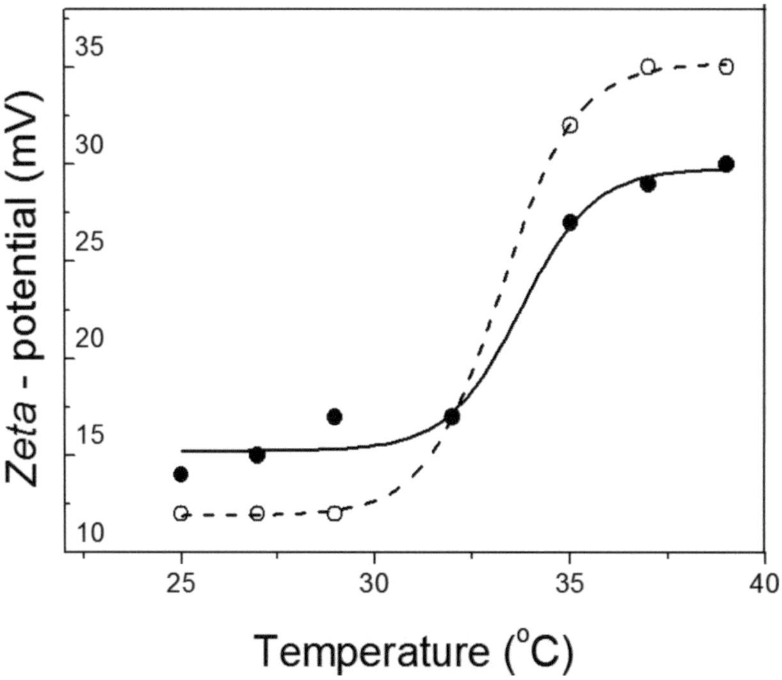


**Figure 10.** Zeta-potential profile of gold-loaded microgel composite particles (solid points) and pure microgel template (hollow points) in different temperature conditions.

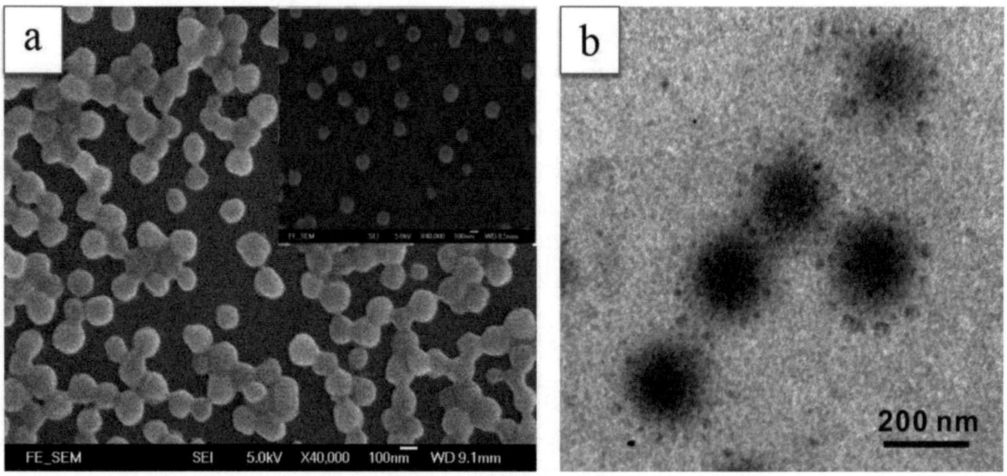


**Figure 11.** (a) SEM image of Au/ (PNIPAm/PEI) composite particles, inset is the original microgel template, (b) TEM image of Au/ (PNIPAm/PEI) composite particles. The particles were synthesized at 25°C and pH 7.30.

### 4.3. UV-vis spectroscopy

The formation of gold nanoparticles in the presence of microgel template was monitored by a UV-vis spectroscopy as a function of time. In this case, gold nanoparticle formation was evident at the absorbance wavelength of 525 nm as shown in **Figure 12**. Gold formation starts after 20 minutes of reaction together with the change in color from turbid white to light pink. An increase of absorbance happens as further reaction occurs for gold nanoparticle formation. This also gives rise on the concentration of gold nanoparticle at higher absorbance. After 4 hours of reaction, it was observed that there were no more significant changes in the absorbance intensity, which signify that gold nanoparticles have ceased to grow or gold ions have ultimately been reduced to nanoparticles. Wavelength absorbance of gold nanoparticles ranging from 520 to 525 nm is a characteristic of Au nanoparticles with spherical shape with sizes ranging from 15 to 30 nm [74].

### 4.4. High-resolution TEM (HRTEM) and X-ray diffraction

To get a closer look at the image of AuNPs immobilized within the PNIPAm/PEI microgels template, an HRTEM analysis was performed as shown in **Figure 13a**. This image reveals a five-fold twinned Au nanocrystal with a diameter of 22.5 nm. Top inset of **Figure 13a** displays the selected-area of electron diffraction (SAED) pattern of AuNPs examined, which reveals ring patterns indexed as (111), (200), and (222) of a face-centered cubic (FCC) gold lattice. Furthermore, this five-fold twinned boundary at the center of an Au nanocrystal can suggest formation of a multiply twinned particle (MTP) close to an icosahedral gold nanostructure. The fuzzy portion observed on TEM image is attributed to the composite particle's sensitivity to misorientation and distortion of the ideal icosahedrons. Aside from the twin boundary observed, the Au nanostructure is mainly composed of (111) planes with a d-spacing of 0.236 nm as shown in **Figure 13b**. The lattice plane is separated by a twin boundary indicated as a white line on the image. The crystallinity of AuNPs embedded in the microgel template is

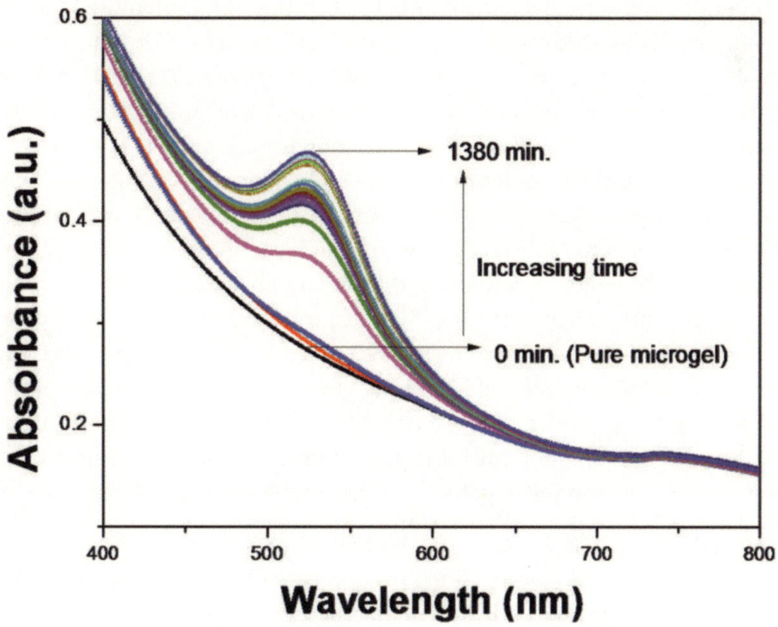


**Figure 12.** UV-vis spectra profile for reduction of gold ions to nanoparticles using a PNIPAm/PEI microgels versus time (minute). Experiment was performed at 25°C, pH 5.6, 200 rpm, with N/Au molar ratio of 28.5.

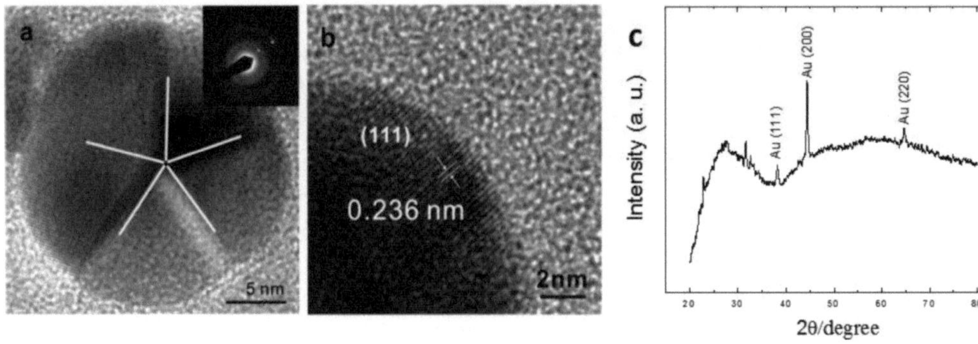


**Figure 13.** (a) HRTEM image of Au nanoparticle embedded within a microgel template, inset is the selected-area electron diffraction (SAED) pattern, (b) (111) planes of Au nanocrystal with a *d*-spacing of 0.236 nm, and (c) XRD spectra of Au/PNIPAm/PEI composite particles.

analyzed through an X-ray diffractometer in **Figure 13c**. Au nanocrystals formed have lattice arrangements of (111), (200), and (220) at corresponding angles of 38, 44, and 65o. This result is consistent with the previous SAED analysis except for the (220) lattice with a dominant (111) arrangement.

### 4.5. Surface composition using X-ray photoelectron spectroscopy

To further investigate the formation of AuNPs using a microgel template, X-ray photoelectron spectroscopy (XPS) analysis at a depth of 10 nm was used. Results shown in **Figure 14** reveal an XPS spectra with binding energies of different elements present in the composite particles. The binding energies correspond to elements of C, O, N, and Au. Convoluted C1s spectra were fitted with peaks at 285.0 and 287.9 eV, assigned to C-C/C-H and C-O bonds, respectively. N1s peak fitted at 399.3 assigned to amines coordinated with AuNPs. O1s at 531.2 eV corresponds to the carbonyl functional group of the microgel. Zero-valent AuNPs are observed from its two peak characteristics at 84.3 and 88 eV, consistent with literature [75]. Other characteristic of AuNP is its XPS spectra peak-to-peak distance of 3.7 eV on the Au 4f doublet which further gives a standard measure of the $Au^0$ oxidation state [76]. As the AuNPs attached to the amino groups, detection of N1s in the XPS analysis weakens due to the overlapping of AuNPs on the amine group [77], eventually strengthening the Au signal. This amine-gold interaction was also observed in the work of Kumar et al. [78] and similar to Manna et al. [79]. Nitrogen peak was also curve fitted into two components at 399.3 and 401.2 eV. The first one corresponds to the amine (free and coordinated to gold) and the other corresponds to the protonated amine or ammonium. With the presence of these two peaks, AuNP binding to the amine group is more on metal-ligand coordination (metallic gold atom and amine) than electrostatic interaction (between ammonium and the negative charges on the surface of the particles).

Further study on the same XPS spectra indicates that there was no significant change of O1s to C1s ratio before and after gold loading. Such insignificant change further proves the presence of the carbonyl functional group, PNIPAm, in the microgel. The XPS spectra also mean that at a depth of at most 10 nm, PNIPAm is present within the shell region partially overlapping the

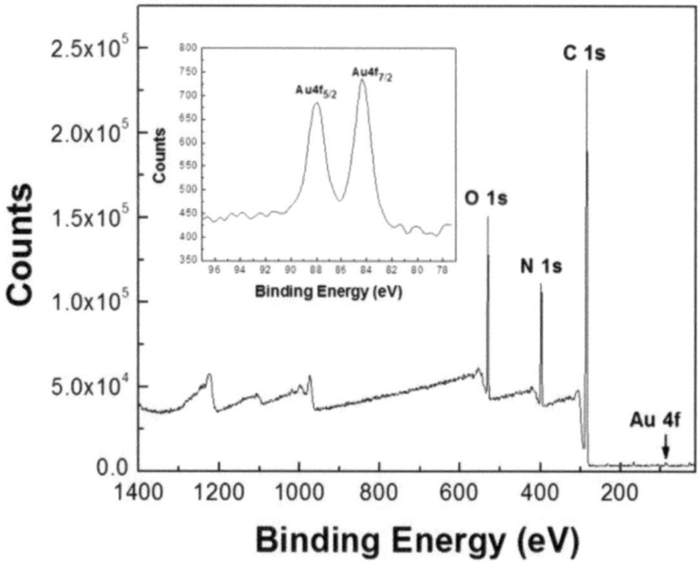


**Figure 14.** XPS spectra of AuNP embedded in PNIPAm/PEI microgel template. Inset is the Au 4f core-level spectra.

core. With this overlapping of shell component to the core makes the whole microgel system shrink whenever PNIPAm shell becomes sensitive to temperature. Such phenomenon further verifies our claim that the amine group residing in the PEI shell l is mainly responsible for the formation and binding of the AuNPs.

## 5. Catalytic activities of gold/microgel and gold@silver/microgel nanocomposite particles

To demonstrate the catalytic activity of gold and gold@silver nanoparticles in a microgel template, a catalytic reduction first order kinetic model (i.e., *p*-nitrophenol reduction by sodium borohydride) was chosen. Silver and gold metal nanoparticles have a wide absorption band in the visible region of the electromagnetic spectrum. Thus, they are easy to characterize and with a wide availability of related literature. These metal nanoparticles have also been involved in many catalytic organic reactions and synthesis in both pure and alloyed form. Previous study suggests that silver preserves the overall spherical morphology of the resultant bimetallic eventually prevents the tendency to phase segregate [85]. These are the primary reasons why these two metals have been chosen to demonstrate the metal nanoparticle forming capabilities of the microgel (i.e., PNIPAm/PEI) template. The *p*-nitrophenol solution exhibits a typical absorption peak at around 320 nm under neutral or acidic condition. When sufficient amount of $NaBH_4$ is added, the nitrophenolate ions become the dominant species and reduce to aminophenol. Such conversion causes the absorption peak to shift to 400 nm. In the absence of any catalyst, the reduction of *p*-nitrophenol by $NaBH_4$

cannot proceed based on a control experiment. And theoretically, this is because the $E_0$ value for the reduction of *p*-nitrophenol to *p*-aminophenol was −0.76 V and that of $H_3BO_3/BH_4^-$ was −1.33 V versus the standard hydrogen electrode (NHE). However, when a reduction of *p*-nitrophenol starts, a new peak appears at about 310 nm, which corresponded to the typical absorption peak of *p*-aminophenol Physical change on the solution color is also obvious during the reaction [80]. For the case of gold@silver metal nanoparticle as catalyst, it only took 2 minutes to complete the catalytic reaction (**Figure 15a**). When using monometallic Au catalyst, catalytic reactions were completed in 15 minutes and 3.5 minutes using different amine to gold ratios (i.e., 28.2 and 14.09, respectively). The catalytic activity of the bimetallic catalyst is obviously higher than that of the monometallic catalyst using the same template.

The kinetic rate constant, which is proportional to its overall kinetic rate in a first order reaction, was estimated from its slope. The control sample has a rate constant of $5.4 \times 10^{-3}$ $s^{-1}$. However, when monometallic gold nanoparticles (N/Au = 28.20 mole ratio) was used as a catalyst, the reaction proceeded approximately 10 times faster (i.e., with a rate constant of $2.44 \times 10^{-2}$ $s^{-1}$) than without catalysts. Moreover, when bimetallic gold@silver nanoparticles were used as a catalyst, the reaction rate was significantly enhanced. The enhancement in catalytic activity is attributed to the synergistic effects and the flexible design between the two metal nanoparticles [81], in this case gold and silver nanoparticles. The electronic and geometrical properties of the synthesized bimetallic nanoparticles can also affect the catalytic activity. Similar studies suggest that the increase in the number of low coordination number, edge and corner sites can also enhance catalytic activity [82]. Surface science studies conclude that the surface electronic structure can be modified by the interactions between the two kinds of atoms in the bimetallic alloy owing to ligand [83] and strain effects [84]. **Figure 15b** shows comparison of the different catalytic reaction rates constants by plotting ln ($C_t/C_0$) versus

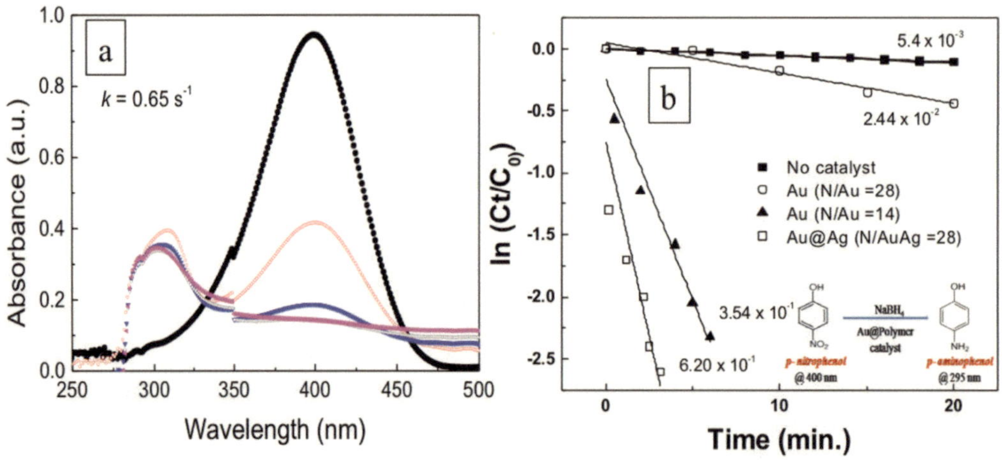


**Figure 15.** (a) UV-vis spectroscopy profile for the reduction of *p*-nitrophenol to *p*-aminophenol using Au@Ag/(PNIPAm/PEI) composite particle as a catalyst. The different colored-curves refer to the different 30 second time intervals, (b) plot of ln ($C_t/C_0$) as a function of time for the reaction catalyzed by Au/PNIPAm/PEI in different N/Au mole ratios and Au@Ag bimetallic nanoparticles in PNIPAm/PEI template. Inset is the reaction scheme of the catalytic reaction model used (i.e., reduction of *p*-nitrophenol to *p*-aminophenol).

http://dx.doi.org/10.5772/intechopen.68388

reaction time for the reduction of *p*-nitrophenol. The results demonstrate that the increase or incorporation of different metal nanoparticles can significantly increase the reduction rate.

### 5.1. Modification of AuNP to Au@Ag bimetallic NP and its effect on catalysis

The main goal in making bimetallic nanoparticles is to enhance the catalytic activity in the reduction of *p*-nitrophenol to *p*-aminophenol. Through the introduction of silver ions into the as-prepared seed gold nanoparticles, surface modification was achieved in the resultant bimetallic Au@Ag nanoparticles. This modification affects the electronic properties of the resultant bimetallic nanoparticles affecting the catalytic activities. Thus, the role of the Ag in the bimetallic structure is a co-catalyst able to promote the ligand effect [85].

Ligand effect suggests that with the presence of a co-catalyst, Ag is important for the redox reaction (i.e., reduction of *p*-nitrophenol) occurring on Ag@Au interfaces [86]. These Ag@Au interfaces are the main actors in improving the catalytic activities. **Figure 16** demonstrates an Ag@Au interface with different work functions (i.e., Au (~5.3 eV) and Ag (~4.7 eV)). Since Ag has a lower work function compared to Au, electrons leave from the Ag atom side of the interface toward the Au side through a depleted region (Region D). As a result, the Au becomes an electron-rich region (Region E). The abundance of electrons on the Au side initiates the uptake of electron from the reactants (i.e., *p*-nitrophenol) on top of the usual uptake from the depleted region. Thus, the more interfaces there are the more depletion and surplus of electron exist, resulting to increase the adsorption of reactants to be reduced on top of the interfacial regions. Such mechanism is consistent with the study of Zhang et al. [87] wherein the increasing electronegativity of Au with

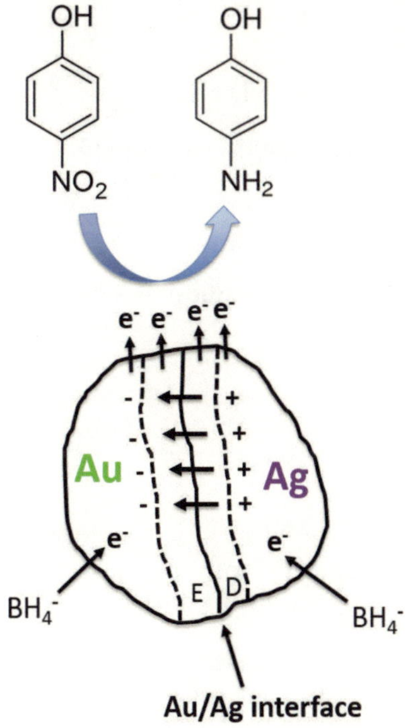


**Figure 16.** Diagram on the transport of electron from an Ag-Au interface bimetallic NP [Adopted from Ref. [87]].

respect to Ag facilitates adsorbate binding, increasing the electron transfer to *p*-nitrophenol. As a result, this reduces the activation energy barrier, thus increasing the catalytic activity.

## 6. Conclusion

Environment-friendly approach on the synthesis of metal/polymeric nanocomposite particles was demonstrated in this chapter through the fabrication of Au and Au@Ag nanoparticles using a microgel template (i.e., PNIPAm/PEI). PNIPAm/PEI microgel template plays a crucial role in the reduction of metal salts, stabilization, and immobilization of the resulting metal/polymer nanocomposites. Furthermore, it can also act as a regulator of metal nanoparticles. Catalytic activities of the Au and Au@Ag metal nanoparticles in microgel template were also demonstrated in the reduction of *p*-nitrophenol to *p*-aminophenol.

## Acknowledgements

We highly acknowledge the Hong Kong Polytechnic University that has given us the chance to explore our interests in Applied Chemistry. And we are grateful to everyone who inspired us to write this chapter.

## Author details

Noel Peter Bengzon Tan[1]* and Cheng Hao Lee[2]

*Address all correspondence to: bengzontan@nami.org.hk

1 Environmental Technologies Section, Nano and Advanced Materials Institute, Ltd., Hong Kong

2 Department of Applied Biology and Chemical Technology, The Hong Kong Polytechnic University, Hong Kong

## References

[1] Li D, Cui Y, Wang K, Hem Q, Yan X, Li J. Thermosensitive nanostructures comprising gold nanoparticles grafted with block copolymers. Advanced Functional Materials. 2007;**17**:3134–3140. DOI: 10.1002/adfm.200700427

[2] Li D, He Q, Li J. Smart core/shell nanocomposites: Intelligent polymers modified gold nanoparticles. Advances in Colloid and Interface Science. 2009;**149**:28–38. DOI: 10.1016/j.cis.2008.12.007

[3] Welsch N, Ballauff M, Lu Y. Microgels as nanoreactors: Applications in catalysis. Advances in Polymer Science. 2010;**234**:129. DOI: 10.1007/12_2-10_71

[4] Li D, He Q, Cui Y, Wang K, Zhang X, Li J. Thermosensitive copolymer networks modify gold nanoparticles for nanocomposite entrapment. Chemistry – A European Journal. 2007;**13**:2224–2229. DOI: 10.1002/chem.200600839

[5] Faraday M. The bakerian lecture: Experimental relations of gold (and other metals) to light. Philosophical Transactions of the Royal Society. 1857;**147**:145–181. DOI: 10.1098/rstl.1857.0011

[6] Turkevich J, Stevenson PC, Hillier J. A study on the nucleation and growth processes in the synthesis of colloidal gold. Discussions of the Faraday Society. 1951;**11**:55–75. DOI: 10.1039/DF9511100055

[7] Brust M, Walker M, Bethell D, Schiffrin DJ, WHyman RJ. Synthesis of thiol-derivatised gold nanoparticles in a two-phase liquid-liquid system. Journal of the Chemical Society, Chemical. Communications. 1994;1:801–802. DOI: 10.1039/C39940000801

[8] Frens G. Controlled nucleation for the regulation of the particle size in monodisperse gold suspensions. Nature: Physical Science 1973;**241**:20–22

[9] Yonezawa T, Kunitake T. Practical preparation of anionic mercapto ligand-stabilized gold nanoparticles and their immobilization. Colloids and Surfaces A: Physicochemical and Engineering Aspects. 1999;**149**:193–199. DOI: 10.1016/S0927-7757(98)00309-4

[10] Hoshetler MJ, Templeton AC, Murray RW. Dynamics of place-exchange reactions on monolayer-protected gold cluster molecules. Langmuir. 1999;**15**:3782–3789. DOI: 10.1021/la981598f

[11] Tzhayik O, Sawant P, Efrima S, Kovalev E, Klug JT. Xanthate capping of silver, copper, and gold colloids. Langmuir. 2002;**18**:3364–3369. DOI: 10.1021/la015653n

[12] Porter Jr LA, Ji D, Westcott SL, Grape M, Czernuszewicz RS, Halas NJ, Lee TR. Gold and silver nanoparticles functionalized by the adsorption of dialkyl disulfides. Langmuir. 1998;**14**:7378–7386. DOI: 10.1021/la980870i

[13] Resch R, Baur C, Bugacor A, Koel BE, Echternach PM, Madhukar A, Montoya N, Requicha AAG, Will P. Linking and manipulation of gold multinanoparticle structures using dithiols and scanning force microscopy. Journal of Physical Chemistry B. 1999;**103**:3647–3650. DOI: 10.1021/jp984508o

[14] Balasubramanian R, Kim B, Tripp SL, Wang X, Lieberman M, Wei A. Dispersion and stability studies of resorcinarene-encapsulate gold nanoparticles. Langmuir. 2002;**18**: 3676–3681. DOI: 10.1021/1a0156107

[15] Weare WW, Reed SM, Warner MG, Hutchison JE. Improved synthesis of samll (dcore = 1.5 nm) phosphine-stabilized gold nanoparticles. Journal of the American Chemical Society. 2000;**122**:12890–12891. DOI: 10.1021/ja002673n

[16] Heath JR, Brandth L, Leff DV. Synthesis and characterization of hydrophobic, organically-soluble gold nanocrystals functionalized with primary amines. Langmuir. 1996;**12**:4723–4730

[17] Selvakannan PR, Mandal S, Phadtane S, Pasricha R, Sastry M. Capping of gold nanoparticles by the amino acid lysine renders them water-dispersible. Langmuir. 2003;**19**: 3545–3549. DOI: 10.1021/la026906v

[18] Joo SW, Kim WJ, Yoon WS, Choi IS. Adsorption of 4,4'-biphenyl diisocyanide on gold nanoparticle surface investigated by surface-enhanced raman scattering. Journal of Raman Spectroscopy. 2003;**34**:271–275. DOI: 10.1002/jrs.994

[19] Li G, Lauer M, Schulz A, Boettcher C, Li F, Fuhrhop JH. Spherical and planar gold(0) nanoparticles with a rogod gold(I)-anion or a fluid gold(0)-acetone surface. Langmuir. 2003;**19**:6483–6491. DOI: 10.1021/ia0300277

[20] Cheng W, Dong S, Wang E. Iodine-induced gold nanoparticle fusion/fragmentation/aggregation and iodine-linked nanostructured assemblies on a glass substrate. Angewandte Chemie International Edition. 2003;**42**:449–452. DOI: 10.1002/anie.200390136

[21] Reetz MT, Helbig, W. Size-selective synthesis of nanostructured transition metal clusters. Journal of the American Chemical Society B. 1994;**116**:7401–7402. DOI: 10.1021/ja00095a051

[22] Bakrania SD, Rathore GK, Wooldrige MS. An investigation of the thermal decomposition of gold acetate. Journal of Thermal Analysis and Calorimetry. 2009;**95**:117–122. DOI: 10.1007/s10973-00809173-1

[23] Huang WC, Chen YC. Photochemical synthesis of polygonal gold nanoparticles. Journal of Nanoparticle Research. 2008;**10**:697–702. DOI: 10.1007/s11051-007-9293-8

[24] Okitsu K, Mizukoshi Y, Yamamoto TA, Maeda Y, Nagata Y. Sonochemical synthesis of gold nanoparticles on chitosan. Materials Letters. 2007;**61**:3429–3231. DOI: 10.1016/j.matlet.2006.11.090

[25] Amendola V, Polizzi S, Meneghetti M. Laser ablation synthesis of gold nanoparticles in organic solvents. Journal of Physical Chemistry B. 2006;**110**:7232–7237. DOI: 10.1021/jp0605092

[26] Seol SK, Kim D, Jung S, Chang WS, Kim JT. One-step synthesis of PEG-coated gold nanoparticles by rapid microwave heating. Journal of Nanomaterials. 2013;**2013**:531760. DOI: 10.1155/2013/531760

[27] Das SK, Marsili E. A green chemical approach for the synthesis of gold nanoparticles: Characterization and mechanistic aspect. Reviews in Environmental Science and Biotechnology. 2010;**9**:199–204. DOI: 10.1007/s11157-010-9188-5

[28] Lengke MF, Southam G. The effect of thiosulfate-oxidizing bacteria on the stability of the gold-thiosulfate complex. Geochimica et Cosmochimica Acta. 2005;**69**:3759–3772. DOI: 10.1016/j.gca.2005.03.012

[29] Gericke M, Pinches A. Biological synthesis of metal nanoparticles. Hydrometallurgy 2006, **83**, 132–140. DOI:10.1016/j.hydromet1.2006.03.019.

[30] Shukla R, Nune SK, Chanda N, Katti K, Mekapothula S, Kulkarni RR, Welshons WV, Kannan R, Katti K V. Soybeans as a phytochemical reservoir for the production and stabilization of biocompatible gold nanoparticles. Small. 2008;**4**:1425–1436. DOI: 10.1002/smll.200800525

[31] Gardea-Torresdey JL, Parsons JG, Gomez E, Videa P, Troiani HE, Santiago P, Jose-Yacaman M. Formation and growth of Au nanoparticles inside live Alfalfa plants. Nano Letters. 2002;**2**:397–401. DOI: 10.1021/nl015673+

[32] Huang J, Li Q, Sun D, Lu Y, Su Y, Yang X, Wang H, Wang Y, Shao W, He N, Hong J, Chen C. Biosynthesis of silver and gold nanoparticles by novel sundried *Cinnamomum camphora* leaf. Nanotechnology. 2007;**18**:105104. DOI: 10.1088/0957-4484/18/10/105104

[33] Song JY, Jang HK, Kim BS. Biological synthesis of gold nanoparticles using *Magnolia kobus* and *Diopyros kaki* leaf extracts. Process Biochemistry. 2009;**44**:1133–1138. DOI: 10.1016/j.procbio.2009.06.005

[34] Iravani S. Green synthesis of metal nanoparticles using plants. Green Chemistry. 2011;**13**:2638–2650. DOI: 10.1039/C1GC5386B

[35] Chen HJ, Wang YL, Wang YZ, Dong SJ, Wang EK. One-step preparation and characterization of PDDA-protected gold nanoparticles. Polymer. 2006;**47**:763–766. DOI: 10.1016/j.polymer.2005.11.034

[36] Sardar R, Park JW, Shumaker-Parry JS. Polymer-induced synthesis of stable gold and silvernanoparticles and subsequent ligand exchange in water. Langmuir. 2007;**23**: 11883–11889. DOI: 10.1021/la702359g

[37] Han J, Liu Y, Li L, Guo R. Poly (o-phenylenediamine) submicroshpere-supported gold nanocatalyst: Synthesis, characterization, and application in selective oxidation of benzyl alcohol. Langmuir. 2009;**25**:11054–11060. DOI: 10.1021/la901373t

[38] Kuo PL, Chen CC, Jao MW. Effects of polymer micelles of alkylated polyethyleneimines on generation gold nanoparticles. Journal of Physical Chemistry B. 2005;**109**:9445–9450. DOI: 10.1021/jp050136p

[39] Cai LJ, Wang M, Hu Y, Qian DJ, Chen M. Synthesis and mechanistic study of stable water-soluble noble metal nanostructures. Nanotechnology. 2011;**22**:285601. DOI:10.1088/0957-4484/22/28/285601.

[40] Wahio I, Xiong YJ, Yin YD, Xia YN. Reduction by the end groups of poly (vinylpyrrolidone): A new and versatile route to the kinetically controlled synthesis of Ag triangular nanoplates. Advanced Materials. 2006;**18**:1745–1749. DOI: 10.1002/adma.200600675

[41] Huang HH, Ni XP, Loy GL, Chew CH, Tan K L, Loh FC, Deng JF, Xu GQ. Photochemical formation of silver nanoparticle in poly (*N*-vinylpyrrolidone). Langmuir. 1996;**12**:909–912. DOI: 10.1021/la950435d

[42] Suzuki D, Kawaguchim H. Gold nanoparticle localization at the core surface by using thermosensitive core-shell particle as a template. Langmuir. 2005;**21**:12016–12024. DOI: 10.1021/la0516882

[43] Dong Y, Ma Y, Zhai T, Zeng Y, Fu H, Yao J. A novel approach to the construction of core-shell gold-polyaniline nanoparticles. Nanotechnology. 2007;**18**:455603. DOI: 10.1088/0957-4484/18/45/455603

[44] Suzuki D, Kawaguchi H. Hybrid microgels with reversibly changeable multiple brilliant color. Langmuir. 2006;**22**:3818–3822. DOI: 10.1021/la052999f

[45] Liu Y, Feng X, Shen J, Zhu JJ, Hou W. Fabrication of a novel glucose biosensor based on a highly electroactive polystyrene/polyaniline/Au nanocomposite. Journal of Physical Chemistry B. 2008;**112**:9237–9242. DOI: 10.1021/jp801938w

[46] Gorelikov I, Field LM, Kumacheva E. Hybrid microgels photoresponsive in the near-infrared spectral range. Journal of the American Chemical Society. 2004;**126**:15938–15939. DOI: 10.1021/ja0448869

[47] Suzuki D, Kawaguchi H. Modification of gold nanoparticle composite nanostructure using thermosensitive core-shell particle as a template. Langmuir. 2005;**23**:8175–8179. DOI: 10.1021/la0504356

[48] Yuan J, Wunder S, Warmuth F, Lu Y. Spherical polymer brushes with vinlimidazolium-type poly (ionic liquid) chains as supported for metallic nanoparticles. Polymer. 2012;**53**:43–49. DOI: 10.1016/j.polymer.2011.11.031

[49] Alexandridis P. Gold nanoparticle synthesis, morphology control, and stabilization facilitated by functional polymers. Chemical Engineering and Technology. 2011;**34**:15–28. DOI: 10.1002/ceat.201000335

[50] Li W, Jia QX, Wang HL. Facile synthesis of metal nanoparticles using conducting polymer colloids. Polymer. 2006; **47**:23–26. DOI: 10.1016/j.polymer.2005.11.032

[51] Ingert D, Pileni MP. Limitations in producing nanocrystals using reverse micelles as nanoreactors. Advanced Functional Materials. 2001;**11**:136–139. DOI: 10.1002/1616-3028(200104)11:2<136::AID-ADFM136>3.0.CO;2-3

[52] Alvarez MM, Khoury JT, Schaaff TG, Shafigullin MN, Vezmar I, Whetten RL. Optical absorption spectra of nanocrystal gold molecules. Journal of Materials Chemistry B. 1997;**101**:3706–3712. DOI: 10.1021/jp962922n

[53] Nalawade P, Mukherjee T, Kapoor S. Green synthesis of gold nanoparticles using glycerol as a reducing agent. Advances in Nanoparticles. 2013;**2**:78–86. DOI: 10.4236/anp.2013.22014

[54] Meristoudi A, Pispas S. Polymer mediated formation of corona-embedded gold nanoparticles in block polyelectrolyte micelles. Polymer. 2009;**50**:2743–2751. DOI: 10.1016/j.polymer.2009.04.045

[55] Ho KM, Li WY, Wong CH, Li P. Amphiphilic polymeric particles with core-shell nanostructures: emulsion-based syntheses and potential applications. Colloid and Polymer Science. 2010;**288**:1503–1523. DOI: 10.1007/s00396-010-2276-9

[56] Tan NPB, Lee CH, Chen L, Ho KM, Lu Y, Ballauff M, Li P. Facile synthesis of gold/polymer nanocomposite particles using polymeric amine-based particles as dual reductants and templates. Polymer. 2015;**76**:271–279. DOI: 10.1016/j.polymer.2015.09.015

[57] Lala NL, Deivraj TC, Lee JY. Auto-deposition of gold on chemically modified polystyrene beads. Colloids and Surfaces A: Physicochemical and Engineering Aspects. 2005;**269**:119–124. DOI: 10.1016/j.colsurfa.2005.06.073

[58] Tan NPB, Lee CH, Li P. Green synthesis of smart metal/polymer nanocomposite particles and their tuneable catalytic activities. Polymer. 2016;**8**:105. DOI: 10.3990/polym804105

[59] Tian C, Mao B, Wang E, Kuang Z, Song Y, Wang C, Li S. Simple strategy for preparation of core colloids modified with metal nanoparticles. Journal of Physical Chemistry C. 2007;**111**:3651–3657. DOI: 10.1021/jp067077f

[60] Zelewsky AV, Barbosa L, Schlaper CW. Poly(ethyleneimines) as bronsted bases and as ligands for metal ions. Coordination Chemistry Reviews. 1993;**123**:229–246

[61] Mitric R, Burgel C, Burda J, Bonacic-Koutecky V, Fantucci R. Structural properties and reactivity of bimetallic silver-gold clusters. European Physical Journal D. 2003;**24**:41–44. DOI: 10.1140/epjde/e2003-00124-7

[62] Herrero E, Buller LJ, Abruna HD. Underpotential deposition at single crystal surfaces of Au, Pt, Ag and other materials. Chemical Reviews. 2001;**101**:1897–1930. DOI: 10.1021/cr9600363

[63] Wang D, Li Y. One-pot for Au-based hybrid magnetic nanostructures via a noble-metal-induced reduction process. Journal of the American Chemical Society. 2010;**132**:6280–6281. DOI: 10.1021/ja100845v

[64] Sato T, Ruch R. Stabilization of Colloidal Dispersion by Polymer Adsorption. New York: Marcel Dekker Inc.; 1980

[65] Marchetti B, Joseph Y, Bertagnolli H. Amine-capped gold nanoparticles: Reaction steps during the synthesis and the influence of the ligand on the particle size. Journal of Nanoparticle Research. 2011;**13**:3353–3362. DOI: 10.1007/s11051-011-0358-3

[66] Chen CC, Kuo PL, Cheng YC. Spherical aggregates composed of gold nanoparticles. Nanotechnology. 2009;**20**:055603. DOI: 10.1088/0957-4484/20/5/055603

[67] Liu Y, Fan Y, Yuan Y, Chen Y, Cheng F, Jiang SC. Amphiphilic hyperbranched copolymers bearing a hyperbranched core and a dendritic shell as novel stabilizers rendering gold nanoparticles with an unprecedentedly long lifetime in the catalytic reduction of 4-nitrophenol. Journal of Materials Chemistry. 2012;**22**:21173–21182. DOI: 10.1039/C2JM34445A

[68] Dey P, Blakey I, Thurecht KJ, Fredericks PM. Self-assembled hyperbranched polymer-gold nanoparticle hybrids: Understanding the effect of polymer overage on assembly size and SERS performance. Langmuir. 2013;**29**:525–533. DOI: 10.1021/la304034b

[69] Takagishi T, Okuda S, Kuroki N, Kozuka H. Binding of metal ions by polyethyleneimine and its derivatives. Journal of Polymer Science. 1985;**23**:2109–2116. DOI: 10.1002/pol.1985.170230804

[70] Horn D, Goethals EJ, editors. Polymeric Amines and Ammonium Salts. Oxford: Pergamon Press; **1980**. p. 333

[71] Wulff G, Dhal PK. Synthesis and Separations Using Functional Polymers. In: Sherrington DC, Hodge P, editors. Book Review: Zeolites, Crystal Growth, Polymers and Miscellanea. London: Wiley; 1988. p. 325

[72] Akamatsu K, Shimada M, Tsuruoka T, Nawafune H, Fujii S, Nakamura Y. Synthesis of pH-responsive nanocomposite microgels with size-controlled gold nanoparticles from ion-doped, lightly cross-linked poly(vinylpyridine). Langmuir. 2010;**26**:1254–1259. DOI: 10.1021/la902450c

[73] Ou JL, Chang CP, Ou KL, Tseng CC, Ling HW, Ger MD. Uniform polystyrene microsphere decorated with noble metal nanoparticles formed without using extra reducing agent. Colloids and Surfaces A: Physicochemical and Engineering Aspects. 2007;**305**:36–41. DOI: 10.1016/j.colsurfa.2007.04.038

[74] Haiss W, Thanh NTK, Aveyard J, Fernug DG. Determination of size and concentration of gold nanoparticles from UV-Vis spectra. Analytical Chemistry. 2007;**79**:4215–4221. DOI: 10.1021/ac0702084

[75] Leff DV, Brandt L, Heath JR. Synthesis and characterization of hydrophobic, organically-soluble gold nanocrystals functionalized with primary amines. Langmuir. 1996;**12**:4723–4730. DOI: 10.1021/la960445u

[76] Yang X, Shi M, Zhou R, Chen X, Chen H. Blending of $HAuCl_4$ and histidine in aqueous solution: A simple approach to the $Au_{10}$cluster. Nanoscale. 2011;**3**:2596–2601. DOI: 10.1039/C1NR10287G

[77] Zhang F, Srinivisan MP. Multilayered gold-nanoparticle/polyimide composite thin film through layer-by-layer assembly. Langmuir. 2007;**23**:10102–10108. DOI: 10.1021/la0635045

[78] Kumar A, Mandal S, Selvakannan PR, Pasricha R, Mandale AB, Sastry M. Investigation into the interaction between surface-bound alkylamines and gold nanoparticles. Langmuir. 2003;**16**:6277–6282. DOI: 10.1021/la034209c

[79] Manna A., Imae T, Aoi K, Okada M, Yogo T. Synthesis of dendrimer-passivated noble metal nanoparticles in a polar medium: Comparison of size between silver and gold particles. Chemistry of Materials. 2001;**13**:1674–1681. DOI: 10.1021/cm000416b

[80] Zhao C, Nie S, Tang M, Sun S. Polymeric pH-sensitive membranes—A review. Progress in Polymer Science. 2011;**36**:1499–1520. DOI: 10.1016/j.progpolymsci.2011.05.004

[81] Zhang P, Shao C, Zhang Z, Zhang M, Mu J, Guo Z, Liu Y. In situ assembly of well-dispersed Af nanoparticles (AgNPs) on electrospun carbon nanofibers (CNFs) for catalytic reduction of 4-nitrophenol. Nanoscale. 2011;**3**:3357–3363. DOI: 10.1039/C1NR10405E

[82] bin Saiman MI, Brett GL, Tiruvalam R, Forde MM, Sharples K, Thetford A, Jenkins RL, Dimitratos N, Lopez-Sanchez JA, Murphy DM, Bethell D, Willock DJ, Taylor SH, Knight DW, Kiely CJ, Hutchings GJ. Involvement of surface-bound radicals in the oxidation of toluene using supported Au-Pd nanoparticles. Angewandte Chemie-International Edition. 2012;**51**:5981–5985. DOI: 10.1002/anie.201201059

[83] Kesavan L, Tiruvalam R, Ab Rahim MH, bin Saiman MI, Enache DI, Jenkins RL, Dimitratos N, Lopez-Sanchez JA, Taylor SH, Knight DW, Kiely CJ, Hutchings GJ. Solvent-free oxidation of primary carbon-hydrogen bonds in toluene using Au-Pd alloy nanoparticles. Science. 2011;**331**:195–199. DOI: 10.1126/science.1198458

[84] Jirkovsky JS, Panas I, Ahlberg E, Halasa M, Romani S, Schiffrin DJ. Single atom hot-spots at Au-Pd nanoalloys for electrocatalytic $H_2O_2$ production. Journal of the American Chemical Society. 2011;**133**:19432–19441. DOI: 10.1021/ja206477z

[85] Garcia AG, Lopes PP, Gomes JF, Pires C, Ferreira EB, Lucena RGM, Gasparotto LHS, Tremiliosi-Filho G. Eco-friendly synthesis of bimetallic AuAg nanoparticles. New Journal of Chemistry. 2014, **38**,2865–2873. DOI:10.1039/c4nj00041b.

[86] Huang J, Vongehr S, Tang S, Lu H, Shen J, Meng X. Ag dendrite-based Au/Ag bimetallic nanostructures with strongly enhanced catalytic activity. Langmuir. 2009;**25**(19): 11890–11896. DOI: 10.1021/la9015383

[87] Zhang H, Haba M, Okumura M, Akita T, Hashimoto S, Toshima N. Novel formation of Ag/Au bimetallic nanoparticles by physical mixture of monometallic nanoparticles in dispersion and their application to catalysts for aerobic glucose oxidation. Langmuir. 2013;**29**:10330–10339. DOI: 10.1021/la401878g

Chapter 6

# Environmental-Friendly Catalytic Oxidation Processes Based on Hierarchical Titanium Silicate Zeolites at SINOPEC

Changjiu Xia, Xinxin Peng, Yao Zhang, Baorong Wang, Min Lin, Bin Zhu, Yibin Luo and Xingtian Shu

Additional information is available at the end of the chapter

http://dx.doi.org/10.5772/intechopen.68389

**Abstract**

Since it was claimed by EniChem in 1983 for the first time, titanium silicate-1 (TS-1) zeolite presented the most delightful catalytic performance in the area of selective organic oxidation reactions. To enhance the mass diffusion property, hierarchical titanium silicate with hollow cavities within crystal was prepared by using a post-synthesis treatment in the presence of organic template, and then, it was commercially produced and employed in many industrial catalytic oxidation processes, such as propylene epoxidation, phenol hydroxylation, and cyclohexanone ammoximation. Moreover, we also developed several totally novel oxidation reactions on hollow titanium silicate (HTS) zeolite, i.e., Baeyer-Villiger oxidation of cyclohexanone and chlorohydrination of allyl chloride with HCl and $H_2O_2$. In all cases, HTS shows much better catalytic performance than TS-1, attributing to the mass diffusion intensification by introducing hollow cavities. On the other hand, enormous works on synthesizing hierarchical TS-1 zeolites with open intracrystalline mesopores have been done via silanization treatment and recrystallization. Based on them, several bulk molecule oxidation processes with *tert*-butyl hydroperoxide, such as epoxidation of fatty acid methyl ester (FAME) and large olefins, have been carried out. As a consequence, hierarchical TS-1 zeolites supply a platform for developing environmental-friendly catalytic oxidation processes to remarkably overcome the drawbacks of traditional routes.

**Keywords:** titanium silicate zeolite, catalytic oxidation, novel chlorohydrination process, HPPO, hierarchical structure, hydrogen peroxide

# 1. Introduction

Framework-incorporated zeolite is an important kind of acid catalyst, which has already been successfully applied or shown great application potentials in many industrial processes in last 30 years [1–4]. Since the first claim of titanium silicate-1 (TS-1) zeolite by EniChem in 1983, serial of metal elements (such as Ti, Sn, Fe, V, Cr, Mn, Zr, Ga, and Ge) have been incorporated into the framework matrix of zeotype materials via the isomorphous substitution of Al atoms in crystalline aluminosilicates through different approaches [5–10]. When $Ti^{4+}$, $Sn^{4+}$, and $Zr^{4+}$ species incorporated into the framework of zeolites, basic subunits are neutral, without needing balancing by cation ions, thus no formation of Brönsted acid sites. However, these metal ions can accept shared electron pairs from electron-rich groups (i.e., peroxide, carbonyl, and C-C bonds) to make substrates much more reactive during catalytic reactions [11–13]. On the other hand, catalytic selective oxidation of hydrocarbons to produce oxygenated organic compounds is of great significance to give birth to chemical intermediates and fuels. Unfortunately, traditional oxidation approaches always suffer from using hazard reagents, complex operation, high energy and cost consumption, low product selectivity, and high pollution [14–16].

To avoid these drawbacks, heterogeneous catalysts (specially based on zeolites) have been introduced to develop highly efficient oxygenated hydrocarbons transformation due to their unique porosity and Lewis acidity for activating oxidants and/or subtracts [17–19]. For example, in conventional route, cyclohexanone oxime is produced via a hydroxylamine method, which needs four steps, involving ammonia oxidation, production of hydroxylamine with liquid acids, cyclohexanone ammoximation, and decomposition of ammonium, generating a large amount of waste and side products [20–22]. In 1986, a totally novel "one-step" cyclohexanone ammoximation process with ammonia ($NH_3$) and hydrogen peroxide ($H_2O_2$) was developed by EniChem via using TS-1 zeolite as a catalyst under mild conditions, with water only as a by-product [23, 24]. Obviously, Ti-incorporated zeolites provide excellent opportunity to update the efficiency of oxidation processes. However, TS-1 zeolite suffered the limitations of mass diffusion, its narrow 10 member-ring channel [25–27]. Thus, a novel post-synthesis route to modify the traditional TS-1 was developed by the researchers at SINOPEC before 2000, obtaining the far more stable hollow titanium silicate (HTS) zeolite. It is worthy to note that HTS zeolite has been commercially manufactured by SINOPEC, and applied in several green catalytic oxidation processes, such as cyclohexanone ammoximation, phenol hydroxylation, and propylene epoxidation [28–30]. Moreover, several completely new reactions, such as chlorohydrination of allyl chloride and Baeyer-Villiger oxidation of cyclohexanone, were initially developed on using HTS zeolite [31, 32]. In recent years, a spongy-like TS-1 zeolite with abundant of open mesopores was prepared via combining the silanization treatment and post-synthesis route under hydrothermal conditions. And it also found some bulk molecule oxidation reactions based on this hierarchical TS-1, such as epoxidation of methyl oleate and large olefins. Thus, this review will majorly summarize the application of hierarchical TS-1 we synthesized in the area of oxidation, as shown in **Figure 1**.

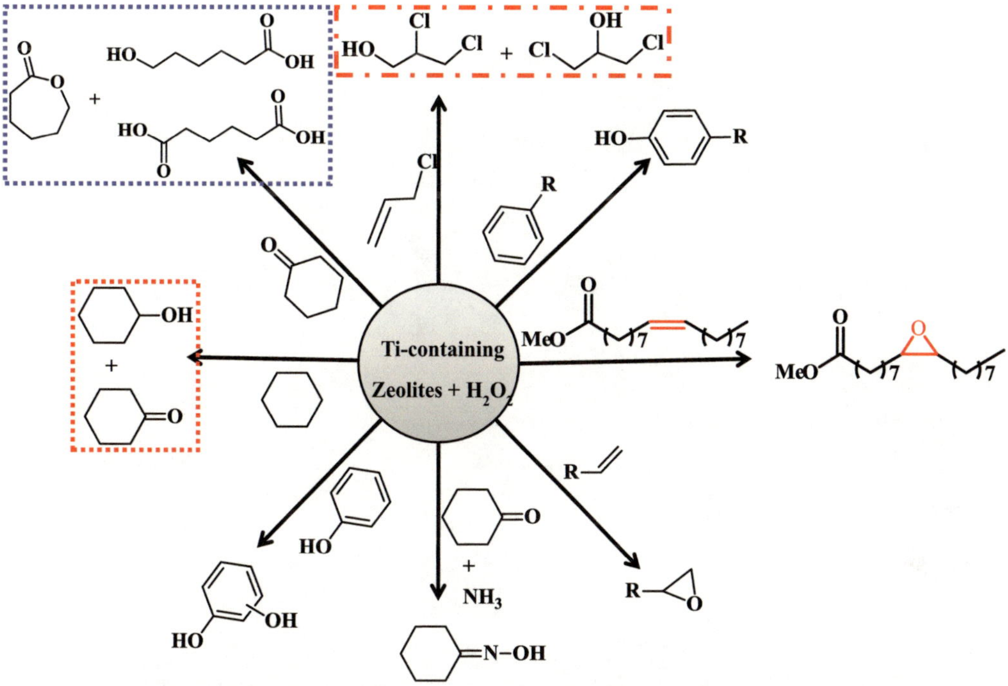


**Figure 1.** Catalytic oxidation reactions on hierarchical titanium silicate zeolite with $H_2O_2$ as an oxidant under mild conditions.

## 2. Synthesis strategies of hierarchical titanium silicate zeolites

TS-1 zeolite was claimed by the researcher EniChem in 1983 for the first time, and then it attracted kind attentions from the world, owing to its pioneering role in exploiting metal-incorporated zeolites [4]. However, in the early years, the key factors for TS-1 zeolite synthesis were far beyond well understanding, thus it was hard to repeat the synthesis of TS-1 in parallel, even under the same conditions. That is ascribed to the different crystallization behaviors between Ti and Si precursors during hydrothermal synthesis processes. To overcome this major drawback, a post-synthesis method, involving the simultaneous dissolution of framework atoms and recrystallization of low aggregated species, was developed by M. Lin and coworkers at SINOPEC in the end of 1990s [33–35]. In a typical procedure, TS-1 was synthesized via the EniChem published method for the first time; and then it was calcined and treated by specific agents; in the end, TS-1 was recrystallized in the presence of SODs at high temperature with autogenous pressure. It has been confirmed that unique intracrystalline cavities (about several to several tens nm in size) are generated through post-treatment, and this material is referred to as hollow titanium silicate (HTS) zeolite, which is of much better diffusion property than TS-1, as shown in **Figure 2** [36]. On the other hand, recrystallization treatment promotes the increasing of framework Ti content, with the significant decreasing of framework defects (Si-OH and Ti-OH groups). Thus, HTS displays far higher stability than

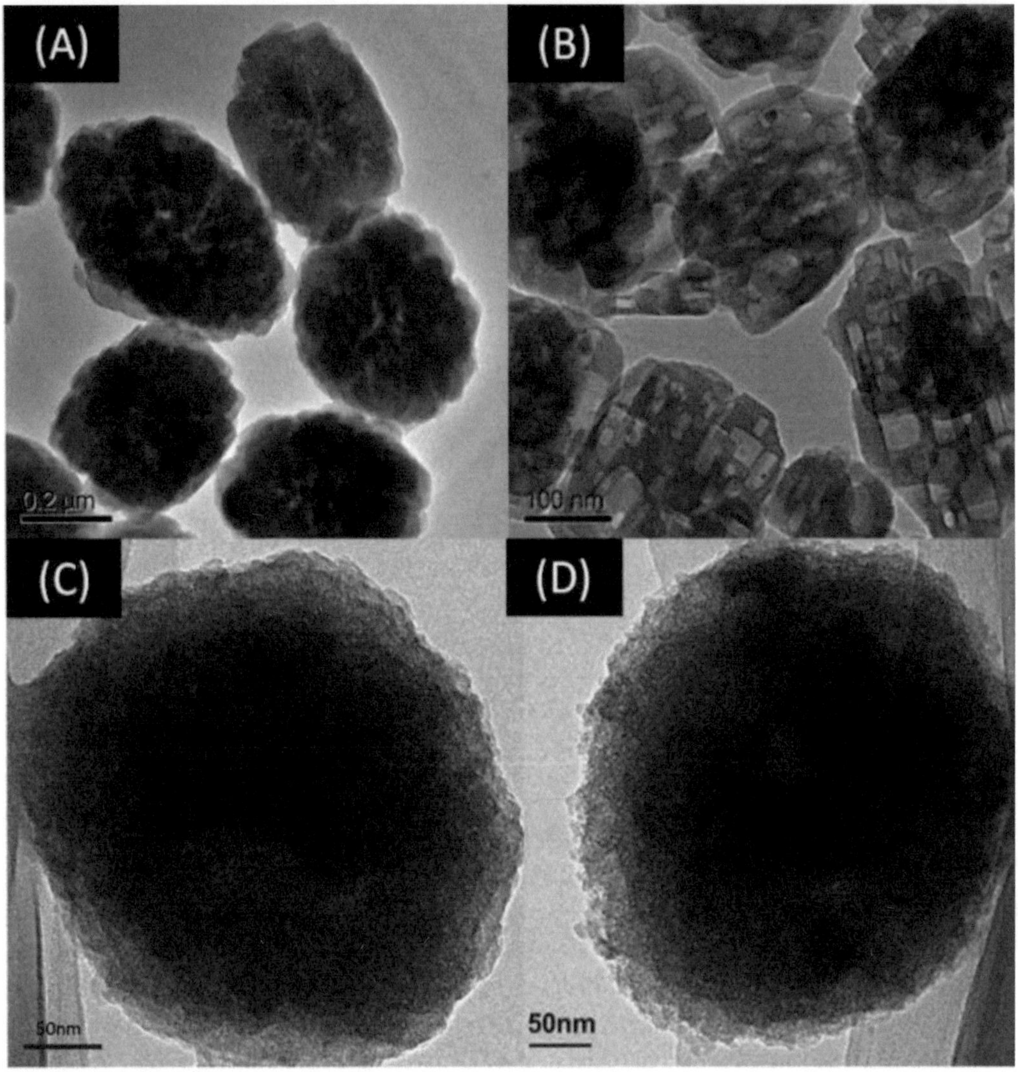


**Figure 2.** Transmission electron microscopy images of TS-1 and hierarchical TS-1 zeolites : (A) TS-1, (B) HTS, (C) TS-OS, and (D) TS-SP.

TS-1 during catalytic oxidation processes, especially in alkaline ammoximation reactions, due to high resistance to ammonia. Up to now, HTS zeolite had been already commercially manufactured at SINOPEC, and successfully applied in many selective catalytic oxidation processes, i.e., cyclohexanone ammoximation, propylene epoxidation, and phenol hydroxylation, presenting desirable economic and social benefits via reducing cost and waste production. To the best of our best knowledge, HTS is the first and only hierarchical titanium-based zeolite applied in industry, and the detailed synthesis, characterization, and application have been summarized in our previous chapter [37].

However, these cavities inside HTS zeolite cannot favor the accessibility of bulk organic molecules to inner tetrahedral framework Ti sites, owing to the 10-member ring pore mouth. To solve this problem, many attempts and efforts have been paid through introducing mesopores and/or macropores with zeolite particles, including selective removing framework-composite atoms, channel filling with soft or hard templates, reassembly of small crystals around mesopore templates, and adding amphiphilic organic surfactants and silicon sources. Among them, hierarchical TS-1 zeolite of rich open mesopores (remarked as TS-OS, as shown in **Figure 2C**) synthesized by introducing amphiphilic organic silane agents is considered as another potential industrial catalyst in the near future [38, 39]. It is very easy and available to synthesis TS-OS even in industry, just adding a few amount of organic silane agents, such as N-[3-(trimethoxysilyl) propyl]aniline and N,N-diethyl-3-(trimethoxysilyl) propan-1-amine, into conventional TS-1 synthesis mixture under hydrothermal conditions. That is attributed to the inhibition of crystal growth by organic silane groups inside particles, as Si-C bonds are ultrastable at high temperature. It is noted that silane groups are easily aggregated, thus mesopores are formed when the organic groups are removed by calcination in air at 550°C. Of course, compared with TS-1, TS-OS is of high mass transfer capability, allowing the entrance of bulk organic subtracts and oxidants. Thus, TS-OS exhibits much better catalytic performance than TS-1 in some specific oxidations. It is worthy to mention that organic silane, usually serving as coupling agents for binder and coating, are very cheap and easily obtained in market, so it is available to synthesis of TS-OS at large scale.

In addition, to make inner mesopores be more uniform, TS-OS zeolite was post-synthesized in the presence of long chain quaternary ammonium hydroxide (such as cetyltrimethylammonium hydroxide, CTAOH) under hydrothermal conditions, to give birth to spongy-like TS-1 zeolite (labeled as TS-SP), as shown in **Figure 2D**. As CTAOH can enter into the crystal of TS-OS along mesopores rather than micropores, the dissolution of framework atoms and reassembly of soluble Ti and Si species under the impact of CTAOH micelles with zeolite particles are simultaneously promoted during post-synthesis processing, thus forming the relatively uniform mesopores, as demonstrated in **Figure 3**. There are two peaks existed for TS-OS, the first one is ascribed to mesopores created by silane agents and the second one is related to intercrystalline mesopore among small crystal particles. The band for TS-SP is very narrow, suggesting the intracrystalline mesopores become much more uniform after reassembly.

## 3. Nature of the activating of $H_2O_2$ molecules by TS-1 zeolite

As it is well known that TS-1 zeolite is an excellent catalyst via improving the nucleophilic attack capability of aqueous $H_2O_2$ solution in catalytic oxidation reactions [40, 41]. When the $H_2O_2$ molecules are adsorbed on the tetrahedral Ti active sites (**Figure 4**), the negative charges on both $O_a$ and $O_b$ atoms in TS2 are −0.460 and −0.476, respectively, which are of much stronger nucleophilic attack capability than that in $H_2O_2$ molecules (about −0.426), as listed in **Table 1**. It indicates that the $H_2O_2$ molecules are activated, with the formation of Ti-OOH species.

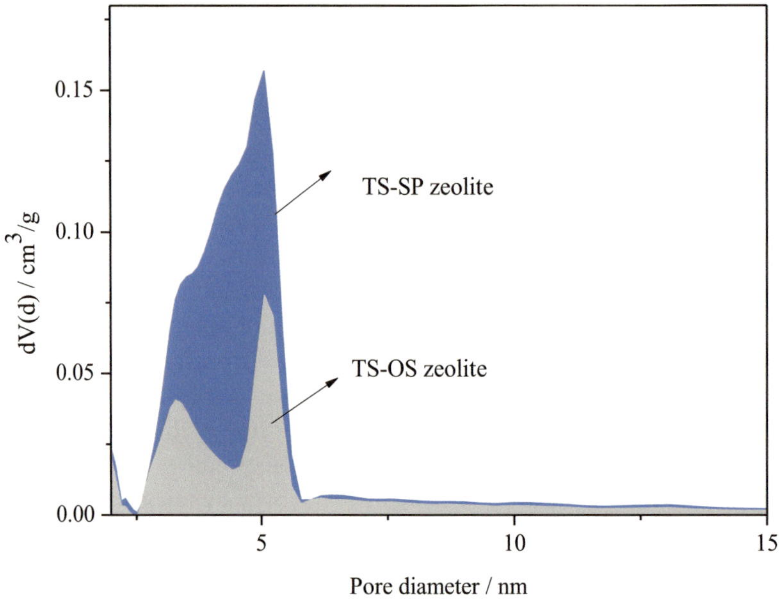


**Figure 3.** Pore-size distribution of TS-OS and TS-SP zeolites calculated by the nonlocal density function theory (NL-DFT) method.

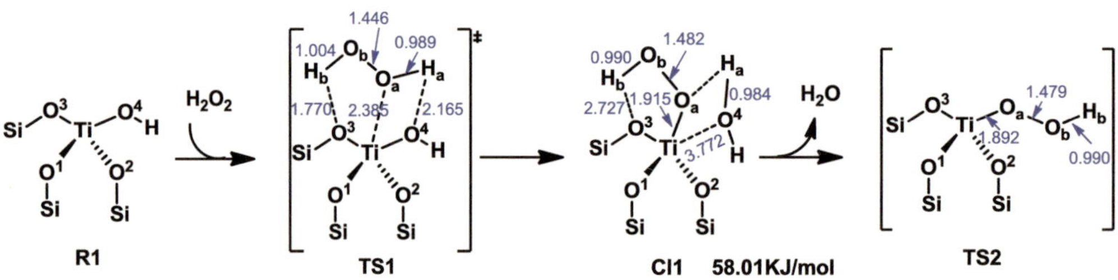


**Figure 4.** Activation of $H_2O_2$ molecules by tetrahedral framework Ti sites calculated by the density function theory (DFT) method.

## 4. Chlorohydrination of allyl chloride with HCl and $H_2O_2$ to produce dichloropropanol

Recently, we have developed a totally novel route to prepare dichloropropanol (DCP) through the chlorohydrination of allyl chloride (AC) with HCl and $H_2O_2$ base on HTS as a catalyst, to replaced $Cl_2$ and $H_2O$ as raw materials in traditional route, as shown in **Scheme 1** [42]; that is, because $Cl_2$ is highly hazard and harmful for environment and human health, and difficult to be stored and transfer, producing a large amount of waste. Most importantly, to avoid the production of trichloropropane (TCP), a great deal of water are added to reduce the concentration of $Cl_2$ and its derivatives (HOCl and HCl), therefore causing the low content of DCP (<4 wt%) in final mixture [43, 44]. Obviously, the tradition $Cl_2$ route goes against the principles

http://dx.doi.org/10.5772/intechopen.68389

| | Atoms | Before $H_2O_2$ adsorption | After $H_2O_2$ adsorption | Charge difference |
|---|---|---|---|---|
| $H_2O_2$ | $O_a$ | −0.426 | −0.460 | −0.034 |
| | $O_b$ | −0.426 | −0.476 | −0.050 |
| | $H_a$ | 0.426 | 0.410 | −0.016 |
| | $H_b$ | 0.426 | 0.469 | 0.043 |

**Table 1.** Charge difference of TS-1 cluster and $H_2O_2$ molecule before and after $H_2O_2$ adsorption.

of green and sustainable chemistry, but HCl is very available and less harmful. It is surprisingly to observe that both AC conversion and DCP selectivity simultaneously over 95% under optimized reaction conditions, and DCP concentration is higher than 20 wt% in aqueous solution. Thus, this novel route demonstrates very excellent industrial potential, due to its high efficiency and low waste produced.

To study the mechanism of this route, different Ti-containing catalysts were evaluated by a novel chlorohydrination reaction under almost the same condition, as shown in **Figure 5**. Among the catalysts we tested, HTS shows the highest AC conversion (over 85%), far higher than that of TS-1 (49%), owing to the improvement of mass diffusion through hierarchically closed mesopores within HTS zeolite particles. At the same time, HTS and TS-1 exhibit much higher selectivity (>88%) of DCP (including 1,3-DCP and 2,3-DCP) than other catalysts, suggesting tetrahedral framework Ti species help the formation of DCP. And the major product catalyzed by TS-1 and HTS is 1,3-DCP, while 2,3-DCP is mainly obtained via the traditional $Cl_2$ route. To explicate this phenomena, ring-opening reaction of epichlorohydrin (ECH) with HCl in aqueous solution had been carried out, with a 1,3-DCP selectivity higher than 98% and

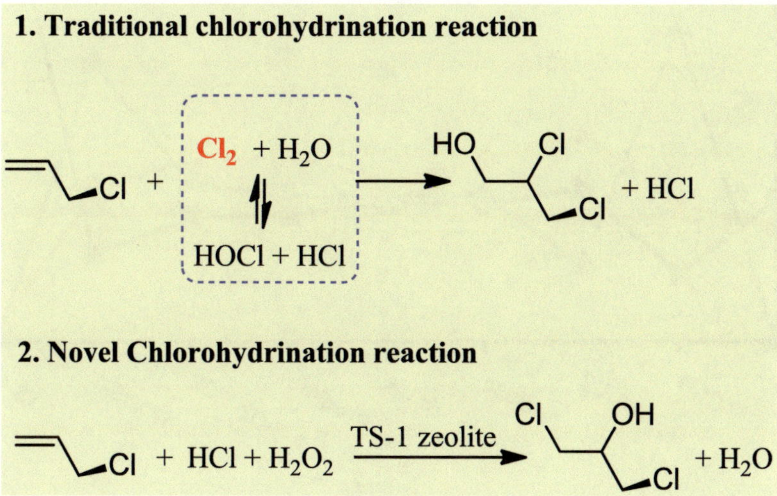


**Scheme 1.** Comparison of traditional and novel chlorohydrination reactions.

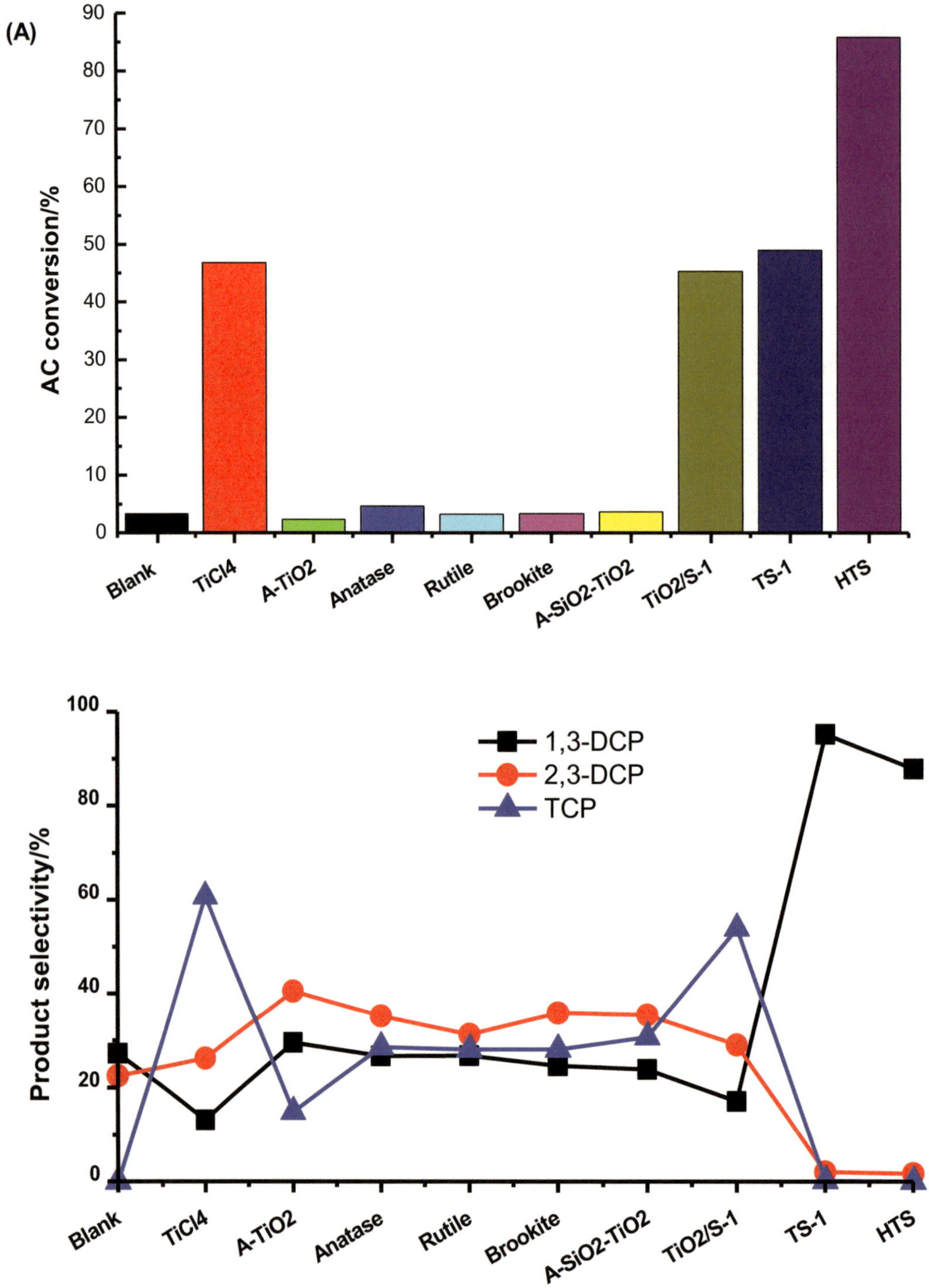


**Figure 5.** Catalytic performance of different Ti-based catalysts in novel chlorohydrination of AC with HCl and $H_2O_2$ under mild conditions: (A) AC conversion and (B) DCP selectivity.

complete AC conversion. In fact, TS-1 zeolite is of high catalytic activity in olefin epoxidation reaction, and many $H_2O_2$ oxidized propylene epoxidation units (called HPPO process) have been established in global. However, TS-1 shows poor catalytic performance in the epoxidation of AC with $H_2O_2$ due to the strong electron-withdrawing effect of Cl atoms on C=C bonds [45, 46]. However, in the presence of HCl, the epoxidation reaction can be accelerated, which is ascribed to the promotion of chemical balance to DCP via ring-opening reaction catalyzed by $H^+$ species.

It is worthy to note that there are still many serious drawbacks related to the HPPO process, attributing to the side-products caused by $H_2O$ [47, 48]. As a result, this route provides a novel method to prepare DCP, with high AC conversion and target product selectivity. Above all, it is indicated that the mechanism of this route lines in the first epoxidation of AC catalyzed by framework Ti species, and then ECH becomes ring opening with HCl. On the other hand, $Cl_2$ and its derivatives (such as $Cl_3^-$, but no HOCl) are detected by UV-Raman spectroscopy (with an irradiation wavelength of 325 nm). It means that the $Cl_2$-based chlorohydrination process can also be occurred in this route, similar as the traditional route, inferring that epoxidation of AC competes with the route of $Cl_2$ [49–52]. What is more, there is only TCP generated catalyzed by HTS and TS-1 zeolites, which is relevant with the $Cl_2$ formed via oxidation of HCl with $H_2O_2$, significantly increasing the utilization efficiency of AC to prepare DCP. Except for the desired catalytic activity and selectivity, HTS also gives much higher stability than TS-1 when they are treated by 37 wt% HCl solution at room temperature. It is found that, after 3 months treatment, HTS zeolite still remains as high AC conversion as fresh one, while the catalytic performance of TS-1 becomes fast decreasing when it was acidic treated for 5 days. As shown in **Figure 6**, HTS can be used six times in a novel chlorohydrination route, without any decreasing of AC conversion and DCP selectivity, which is attributed to the reincorporation of Ti species into framework through post-synthesis treatment.

## 5. Ammoximation of ketones with $H_2O_2$ and $NH_3$

Ammoximation reaction is very important to produce oximes from ketones and aldehydes in industry [53–55]. For example, cyclohexanone oxime, which is a key intermediate for manufacturing ε-caprolactam, is conventionally prepared via the hydroxylamine oxidation route, including hydroxylamine sulfate oxidation (HSO) and hydroxylamine phosphoric oxidation (HPO) approaches [56, 57]. Generally, it involves many harsh drawbacks in these approaches, i.e., long and complex reaction pathways, high energy and material consumed, generating a large amount of waste and side-products, and harmful for health and environment, as shown in **Scheme 2**. To overcome these drawbacks, a "one-step" cyclohexanone ammoximation process with $H_2O_2$ and $NH_3$ was developed by EniChem in 1986 using TS-1 as a catalyst under very moderate conditions [58, 59]. Compared with traditional route, the novel one is highly clean and environmental friendly, with evidently reducing of waste production and capital cost, only water as a side product. The first 12 Kt/a demonstration unit with TS-1 catalyst had been built up by EniChem in 1994, and a commercial 60 Kt/a unit was carried out by Sumitomo in 2003. By using HTS as a catalyst, a 140 Kt/a commercial cyclohexanone

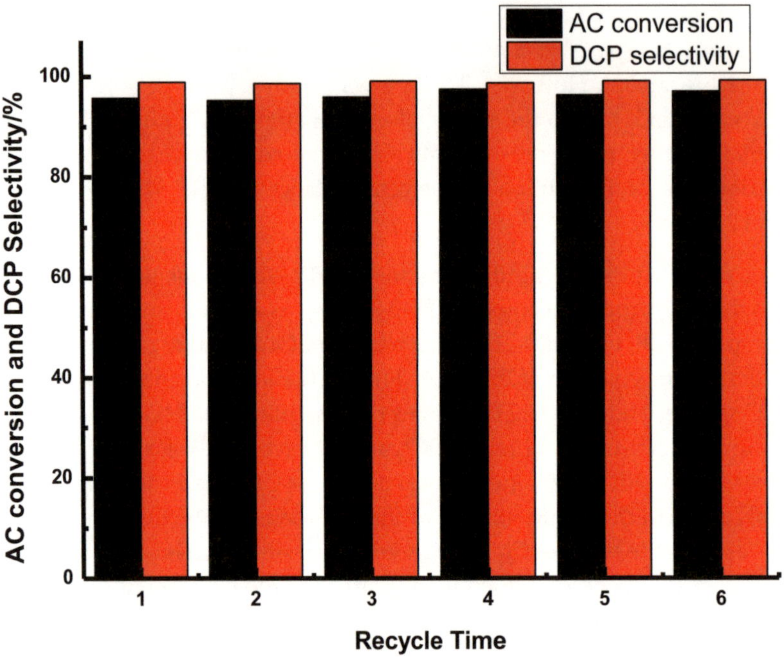


**Figure 6.** Conversion of AC and DCP selectivity in the reuse experiments six times under the same reaction conditions.

ammoximation process was built by SINOPEC, and it run smoothly for over 10 years since 2000 [60]. Similar to this process, acetone ammoximation with $H_2O_2$ and $NH_3$ catalyzed by HTS also carried out, with high acetone conversion and oxime selectivity [61].

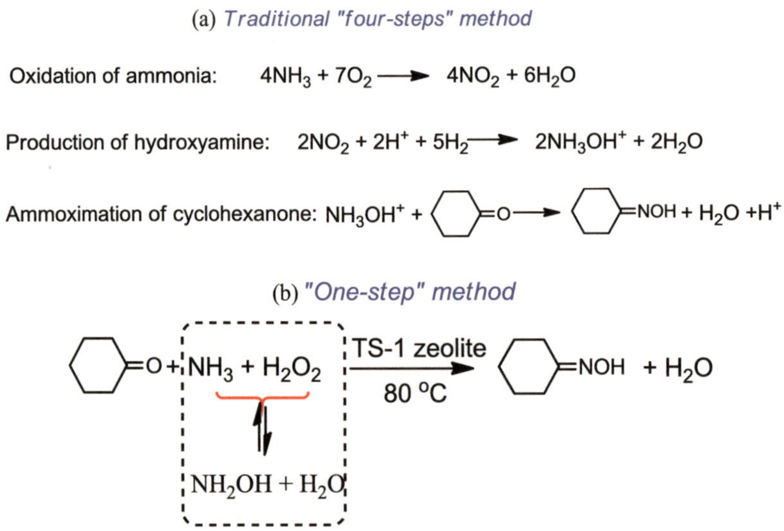


**Scheme 2.** Comparison of traditional HSO (HPO) route and novel TS-1 catalyzed ammoximation reaction for the production of cyclohexanone oxime.

The mechanism of "one-pot" cyclohexanone ammoximation bases on the *in-situ* formation of hydroxylamine ($NH_2OH$) through the oxidation of $NH_3$ with $H_2O_2$ under the effect of tetrahedral Ti species, and then it reacts with cyclohexanone to form oxime [62, 63]. The coordination of $NH_3$ and $H_2O_2$ with framework Ti sites has been investigated by a spectroscopic method, perfectly with an octahedral coordination. And the cyclohexanone ammoximation reaction was operated in a stirred reactor through a continuous liquid phase flow, with a slurry medium (TS-1 catalyst concentration is about 2–3 wt%) at 80–90°C. It is noted that the TS-1 and HTS zeolite particles are about 300–500 nm in size, which are too small to be separated by conventional methods and equipment. Thus, the separation of catalyst from reaction mixture and solvent (*tert*-butanol) is usually carried out in a membrane filter, and then crude oxime is withdrawn from the bottom of this reactor for further purification, and the spent catalyst should be separated from this reaction system and calcined to remove the organic deposit, which fills the channel of TS-1 zeolite, as shown in **Figure 7**.

As shown in **Figure 8**, we can see that HTS zeolite is of far higher catalytic performance and longer lifetime than TS-1 under the same conditions. This is because hollow cavities well enhance the mass diffusion within zeolite particles, and relative short residence time of cyclohexanone favors the decrease of side products, especially for heavy organic deposits. Moreover, the ultrahigh stability of HTS is attributed to the recrystallization of zeolite defects

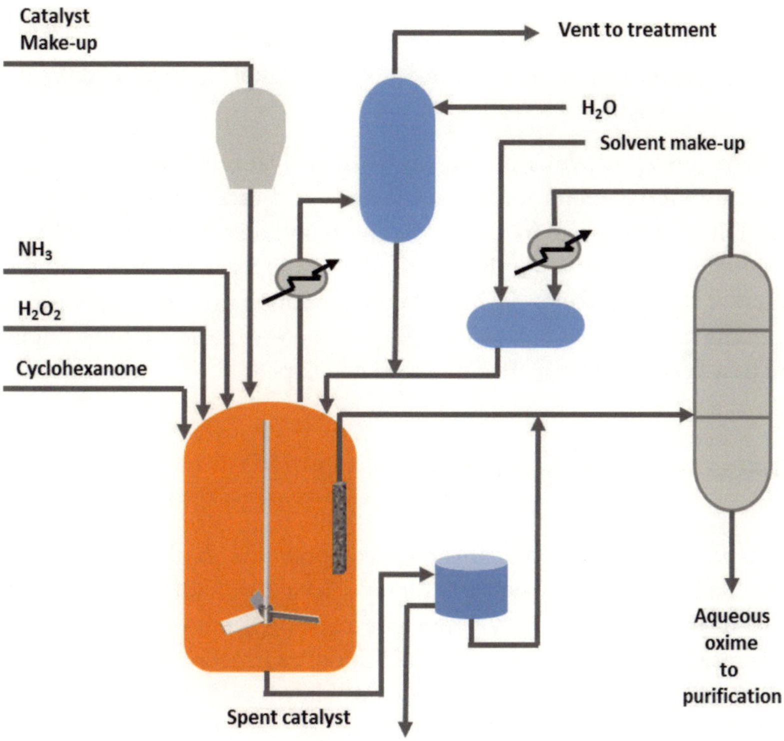


**Figure 7.** Diagrammatic sketch of a "one-step" cyclohexanone ammoximation process catalyzed by TS-1 zeolite under moderate conditions.

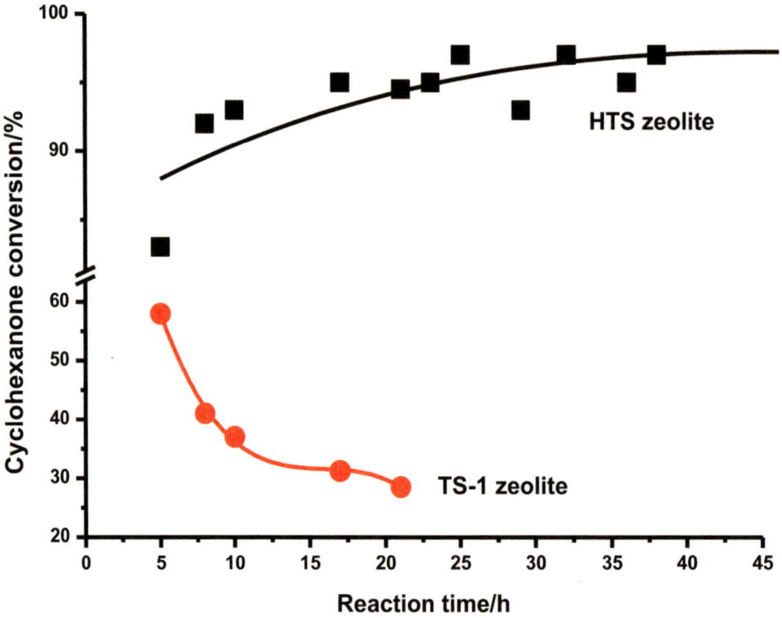


**Figure 8.** Comparison of catalytic performance of HTS and TS-1 zeolites in the cyclohexanone ammoximation process with $H_2O_2$ and $NH_3$.

through dehydrated condensation between Si-OH and Ti-OH groups, making it more tolerant to alkaline intermediates.

In addition, in our previous works, the irreversible deactivation of HTS zeolite has been investigated by using multiple characterization methods and catalytic evaluations [64–66]. It is confirmed that the dissolution of framework Si atoms and the removal of framework Ti atoms cause the formation of acidic amorphous $TiO_2$-$SiO_2$ nanoparticles, leading to the inefficient decomposition of $H_2O_2$ molecules during catalytic oxidation processes. At the same time, the simulated deactivation of HTS in aqueous $NH_3$ solution was also examined, and finally, an efficient and easy regeneration method has been developed.

## 6. Epoxidation of olefins with $H_2O_2$

Olefin epoxidation with $H_2O_2$ to prepare epoxy compound is an ultrakey reaction for manufacturing chemical intermediates, which is highly relevant with all aspects of human life and social production [67–69]. Particularly, propylene oxide (PO), one of the three large propylene derivatives, is majorly produced via a combined chlorohydrination-saponification technology and a cooxidation method in traditional route, as shown in **Scheme 3**. Chlorohydrination route lines in the oxidative addition of propylene with $Cl_2$ and $H_2O$ to give birth to chloropropanol and HCl, and then chloropropanol reacts with $Ca(OH)_2$, with the formation of propylene oxide. We can conclude that the O atom in PO molecule comes from the $H_2O$ molecule, and a large amount of hazardous by-products and wastewater are caused by $Cl_2$ and $Ca(OH)_2$ [70–72]. As a result, a much greener PO manufacture technology catalyzed on TS-1 (remarked

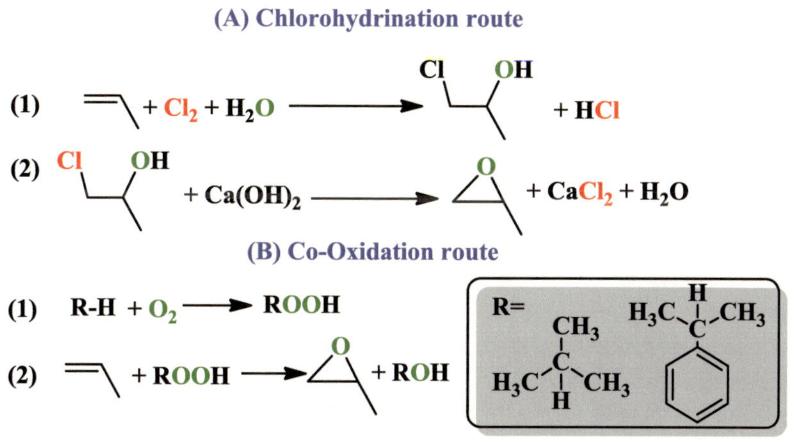


**Scheme 3.** Reaction networks of chlorohydrination route and cooxidation route for the production of propylene oxide in industry.

as HPPO) was exploited by EniChem, and then several commercial HPPO units based on TS-1catalysts were built up and run smoothly in around 2010, significantly reducing the generation of wastewater and unwanted products [73, 74].

**Scheme 4** presents the reaction pathways of epoxidation of propylene with $H_2O_2$ under the impact of tetrahedral Ti species, while the five-member ring (originated from framework

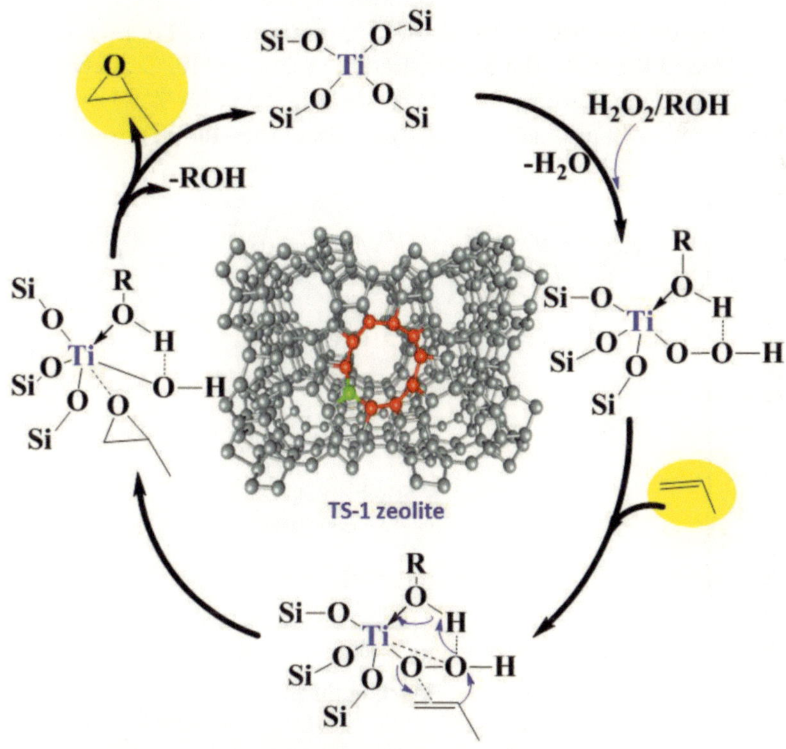


**Scheme 4.** Reaction mechanism of propylene epoxidation with $H_2O_2$ catalyzed by TS-1 zeolite.

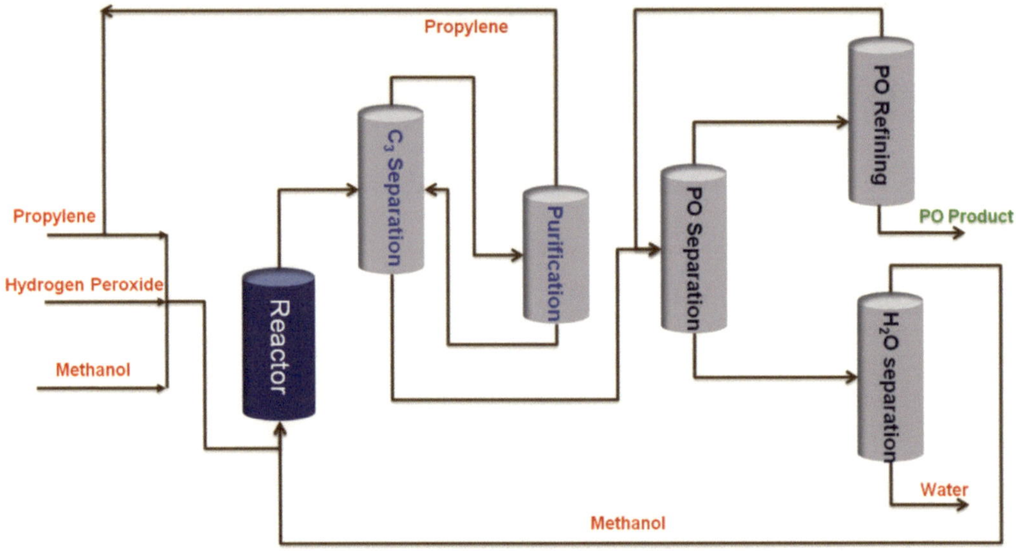


**Figure 9.** Diagrammatic sketch of propylene epoxidation with $H_2O_2$ catalyzed by HTS zeolite.

Ti site, solvent, and $H_2O_2$) mechanism is well accepted, as shown in **Scheme 4**. To verify this mechanism, many characterization and calculation methods were employed and summarized. Since 2000, HTS zeolite was also applied in the HPPO process, showing extreme high catalytic performance and stability. Moreover, the commercial 100 kt/a HPPO unit had been established and run well for over 6000 h, based on the promising result of pilot unit (at a 10 kt/a scale), as illustrated in **Figure 9**. The demonstration process of HPPO catalyzed by HTS was carried out in a fixed-bed reactor under mild conditions, with over 97% $H_2O_2$ conversion and 96–99% PO selectivity at the same time, as shown in **Figure 10**. It is even higher than that

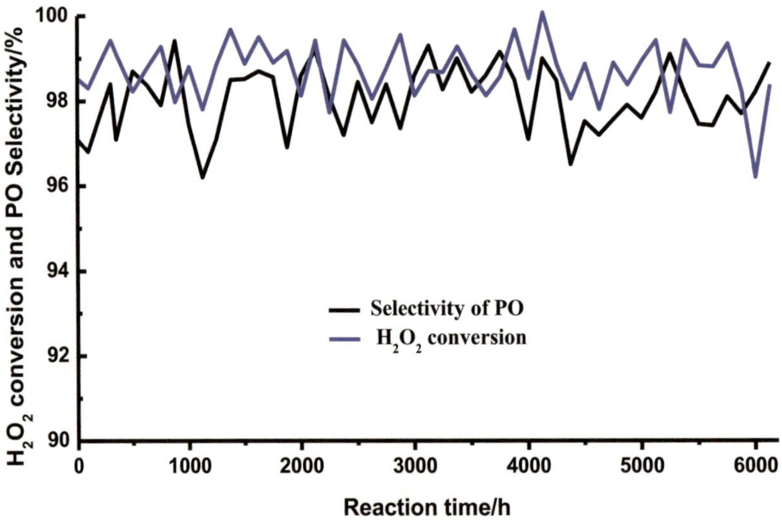


**Figure 10.** Catalytic performance of HTS zeolite in propylene epoxidation at pilot scale.

reported by BASF and DOW (the selectivity of PO is about 93.2%), attributing to the benefits of intracrystalline mesopores [75].

In addition, after the purification of crude PO in a distillation column, the final PO product has a high purity of 99.97%, which had already been used as chemical intermediates to manufacture 1,2-ethyl glycerol, oilfield demulsifier, propylene carbonate, triisopropanolamine, hydroxypropyl, dimethyl carbonate in market. Up to date, one even larger scaled (300 kt/a) industrial HPPO demonstration unit is under constructing, based on the previous experiment results [76].

## 7. Hydroxylation of aromatics with $H_2O_2$

Green hydroxylation of aromatics to phenolic derivatives, such as phenol, cresol, catechol, and hydroquinone, is still a huge challenge for producing many kinds of fine chemicals in industry, due to the strong charge delocalization effect of six-member ring of benzene [77–79]. The production of phenol is mainly on the basis of the commercial cumene process, which contains a three-step process (that is alkylation of benzene with propylene, cumene oxidation, and acid-catalyzed decomposition), with acetone as the major by-product. As acetone is very cheap and superfluous than its consumption, this process is suffered by the capital cost and acetone market. Moreover, this route has a very complex and long pathway, and gives birth to a large amount of by-products due to the severe reaction conditions. As a result, several selective catalysts were proposed, such as V, Fe, and Cu-based catalysts, but they are far beyond to be used in a large scale due to their poor catalytic activity or stability. Direct benzene oxidation to phenol with $H_2O_2$ catalyzed by TS-1 is widely accepted as an alternative process to the traditional cumene process for producing phenol, as shown in **Scheme 5**. As previously mentioned, the function of TS-1 catalyst involves the activation of $H_2O_2$ oxidant to form a highly reactive Ti-OOH species, which can easily nucleophilic attack the C atoms in

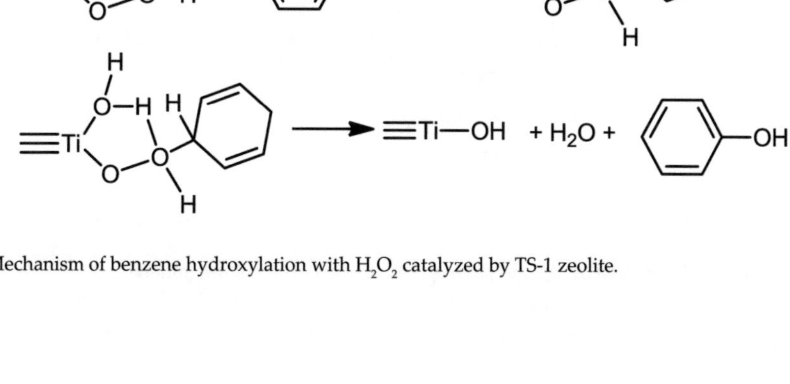


**Scheme 5.** Mechanism of benzene hydroxylation with $H_2O_2$ catalyzed by TS-1 zeolite.

benzene molecules. Moreover, TS-1 zeolite is ultrastable and can be even used in the strong acid media. Some other metal-incorporated zeolites, such as Ti-beta, Ti-MOR, and V-beta, were also evaluated by direct benzene oxidation, but they were of lower catalytic performance than TS-1.

Furthermore, phenol hydroxylation to produce catechol and hydroquinone is also of great importance for fine chemical production, such as agrichemical, photography chemicals, antioxidants, polymerization inhibitors, etc [80, 81]. In conventional routes, the dihydroxybenzenes are commercially produced by several complex processes, including (i) the oxidations of aniline and diisopropylbenzene and (ii) sulfurization of benzene and hydrolysis of benzene sulfonate [82]. **Table 2** shows the detailed technologies for the production of dihydroxybenzenes in industry, but some of them have been sifted out, owing to inevitable limitations. We can see that HTS zeolite demonstrates the highest catalytic performance and technical advantages among these catalysts, especially when taking the environmental and cost benefits into consideration. It was developed and commercially scaled up by EniChem in 1986 for the first time to substitute the Brichima process, avoiding the separation problem of homogeneous catalyst.

After 2000, phenol hydroxylation catalyzed by HTS zeolite was established, with significant reduction of financial investment and operation cost. It is worthy to note that the catalytic activity of TS-1 zeolite in this reaction is closely associated with the chemical status of Ti species; only tetrahedral framework Ti species are able to promote the activation of $H_2O_2$ oxidant, rather than the substrate. Therefore, phenol hydroxylation is usually chosen as a probe reaction to reveal the intrinsic activity of framework Ti sites incorporated.

Additionally, to illustrate the preponderance of hierarchical structure, both TS-1 and HTS zeolites were evaluated by several aromatic (including benzene, toluene and phenol) oxidation reactions, as shown in **Figure 11**. It is observed that HTS zeolite exhibits much higher catalytic performance than TS-1 zeolite in all aromatic oxidation we tested, with ultralong lifetime in these reactions as well. This ascribes to the higher content of framework Ti species of HTS than TS-1, and the improvement of mass diffusion via the formation of hollow cavities, which is in great agreement with the dissolution-recrystallization process through post-synthesis treatment. Accordingly, these results suggest that phenol molecules are more active than toluene, and far more reactive than benzene, in TS-1 catalyzed aromatic oxidations, due to the change of charge property by introducing electronic groups. It is interesting to find that the solvent effect takes great impact on the catalytic performance of TS-1 zeolite, while acetic acid can promisingly enhance toluene conversion in the presence of $H^+$ ions.

| Approach | Catalyst | $X_{phenol}$ (%) | $S_{p\text{- and }o\text{-product}}$ (%) |
|---|---|---|---|
| Rhone-Poulenc | $H_3PO_4/HClO_4$ | 5 | 90 |
| Ube | Ketone peroxide/$H_2SO_4$ | <5 | 90 |
| Brichima | Fe(II)/Co(II) | 10 | 80 |
| EniChem* ($H_2O_2$/phenol = 1/3) | HTS | 26 | >95 |

**Table 2.** Different phenol hydroxylation approaches to produce catechol and hydroquinone.

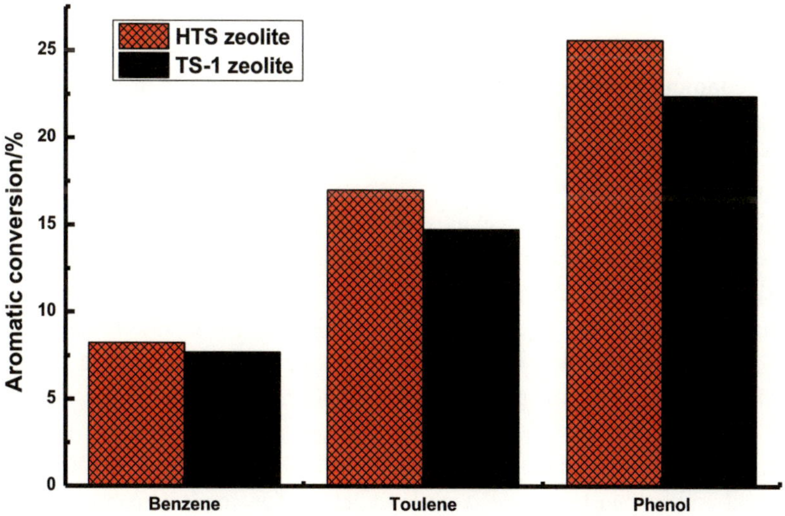


**Figure 11.** Catalytic performance of TS-1 and HTS in various aromatic oxidation with $H_2O_2$, reaction conditions: (A) benzene/$H_2O_2$ = 1:1, solvent free, 70°C, 400 rmp; (B) toluene/$H_2O_2$ = 1:1, acetic acid as solvent, 70°C, 400 rmp; (C) phenol/$H_2O_2$ = 3:1, acetone as solvent, 80°C, 400 rmp.

Generally, the ratio of catechol to hydroquinone is close to 1:1 in the TS-1 catalyzed phenol hydroxylation route, and catechol (para-dihydroxybenzene) is more expansive and of even larger requirement than the other one. Recently, we have proposed an efficient route to modify the product distribution, with obviously increasing of para-product selectivity, by introducing acidic-basic sites via adding MgO-$Al_2O_3$ binary oxides. For MgO-$Al_2O_3$ modified HTS zeolite [83], the optimized ratio of para to ortho product is larger than 2, while the phenol conversion still remains high (>16%). The high *p*-dihydroxybenzene selectivity is ascribed to the steric hindrance and the synergistic effect between Lewis basic-acidic sites of mixed oxides and tetrahedral framework Ti species in HTS zeolite. In other words, HTS zeolite provides an ideal platform to transform aromatics into phenolic derivatives under mild conditions, to make these processes gradually be much cleaner and greener.

## 8. Baeyer-Villiger oxidation of cyclohexanone with $H_2O_2$

Baeyer-Villiger (BV) oxidation is of great importance to convert ketones to esters or lactones, which are important organic intermediates in the fine chemical industry [84–90]. ε-carprolactone is traditionally produced from the BV oxidation reaction of cyclohexanone by using peracids as a catalyst and oxidant, e.g., perbenzoic acid, *m*-chloroperbenzoic acid, and trifluoroperacetic acid. Although peracids route favors high ketone conversion and fast product transformation, it suffers serious pollution and operation drawbacks at commercial scale [91–97]. Aqueous hydrogen peroxide solutions (usually with 30–50 wt% concentration), environmental friendly, cheap, and easy to handle, are considered as a desirable oxidant to replace hazardous peracids. However, hydrogen peroxide is usually of less attacking nucleophilic capability than peracids for activating the carbonyl group of ketones [98–102].

In previous works, much effort has been devoted to develop both homogeneous and heterogeneous catalysts in order to improve the performance of BV oxidation of cyclohexanone [104–111]. One famous catalyst is Sn-beta zeolite, reported by A. Corma, with almost 100% ε-carprolactone selectivity in direct BV oxidation of cyclohexanone in organic solvent at low temperature (<90°C) [112]. It is confirmed that this route bases on the activation of carbonyl group by tetrahedral framework Sn species via the acceptor-donator of electron pairs between framework Sn sites and O atoms in cyclohexanone. In contract, we have found that HTS zeolite is also an effective catalyst for the BV oxidation of cyclohexanone with $H_2O_2$, but further hydrolysis and oxidation of lactone have been occurred with the formation of 6-hydroxyhexanoic acid and adipic acid, respectively. As for the mechanism of this reaction, the density function theory (DFT) method was employed, which suggests that the activation of $H_2O_2$ to form highly reactive Ti-OOH species is the driving force of BV oxidation, as shown in **Figure 12**. However, due to its seven-member-ring structure, ε-carprolactone easily reacts with $H_2O$ to make ring-opening reaction occurred with the formation of 6-hydroxyhexanoic acid. Then, the hydroxyl-substituted acid will further be oxidized by Ti-OOH species, to give birth to adipic acid in the end, as shown in **Figure 13**. To support our proposed mechanism, the major products of cyclohexanone BV oxidation were analyzed by the gas chromatography-mass spectrometry method,

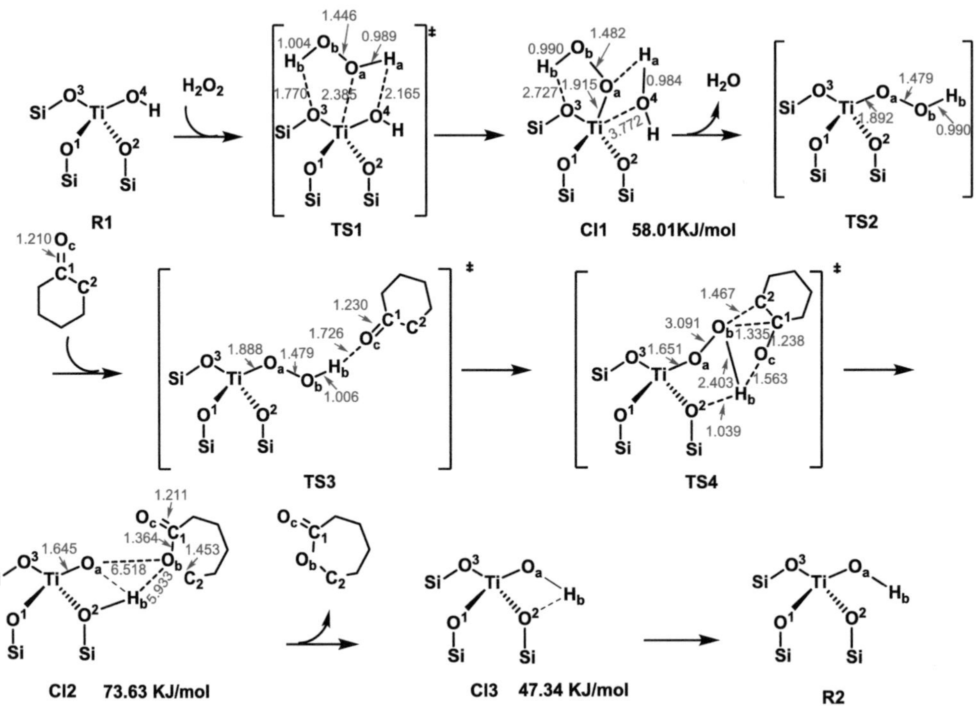


**Figure 12.** One possible reaction pathway of cyclohexanone BV oxidation catalyzed by TS-1 zeolite involving the activation of $H_2O_2$ molecules [103].

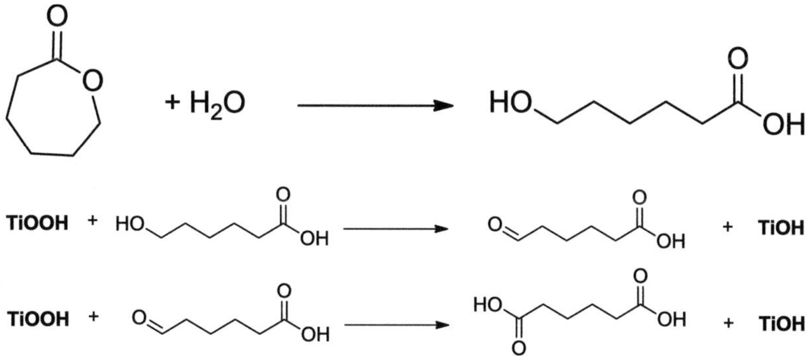


**Figure 13.** Reaction network of ring opening of ε-carprolactone and deep oxidation of 6-hydroxyhexanoic acid to adipic acid.

| Catalyst | Reaction condition | | | Conversion (%) | Target product selectivity (%) | | |
|---|---|---|---|---|---|---|---|
| | Solvent | Temp. (°C) | Time (h) | | 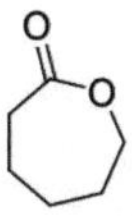 **ε-Carprolactone** | 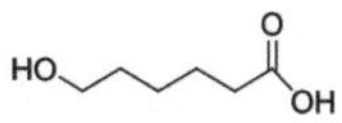 **6-Hydroxyhexanoic acid** | 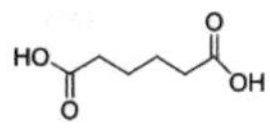 **Adipic acid** |
| HTS | – | 90 | 8 | 59.51 | 0.75 | 61.21 | 28.52 |
| Sn-beta | MTBE/ dioxane | 90 | 3 | 52 | >98 | – | – |

**Table 3.** Reaction parameters and catalytic performance of Sn-beta and HTS in the cyclohexanone BV oxidation with $H_2O_2$.

including 6-hydroxyhexanoic acid, adipic acid, cyclohexane-diol, and cyclohexane-dione. As shown in **Table 3**, HTS zeolite has a much higher conversion (with longer reaction time) than Sn-beta zeolite, without adding any organic solvent. And we also observe that increasing reaction temperature and $H_2O$ content are in favor of high cyclohexanone conversion and acid yield. It is in good agreement with the DFT result which indicated that the framework Ti species display a totally novel catalytic behavior from that Sn species incorporated, attributing to the basic difference of charge property between them.

From the viewpoint of commercial application, although ε-carprolactone selectivity in HTS catalyzed route is far lower, it provides an ideal and green approach to produce adipic acid, overcoming the significant drawbacks of oxidation of KA oil with $HNO_3$, and there is no organic solvent added, reducing the energy consumption for solvent recycling and purification. Most importantly, the product distribution can be altered, according to market requirement, via changing reaction parameters and processing technologies.

## 9. Catalytic oxidation of cyclohexane with $H_2O_2$ to KA oil

Cyclohexane oxidation to prepare cyclohexanone and cyclohexanol (referred as to KA oil) is always a hot spot in the organic chemical industry, contributing to the wide application area of KA oil [113–115]. Currently, KA oil is mainly manufactured by the air oxidation of cyclohexane via the formation and decomposition of cyclohexyl peroxide at over 100°C in industry, using soluble cobalt carboxylate and metal-boric acid as a catalyst, or without a catalyst, with about 80% selectivity of KA oil, as shown in **Scheme 6**. However, the cyclohexane conversion should be kept at ultralow level, about 4–10% for single transformation, to reducing the selectivity of undesired products, such as adipic acid and deep hydroxylated products. Apart from the harsh reaction parameters, homogeneous catalysts are usually difficult to be separated from mixture, with high power consumption. During the past centuries, huge efforts on developing benign oxidation of cyclohexane with various oxidants (including $O_2$, $H_2O_2$, and *tert*-butyl hydroperoxide (TBHP) and catalysts have been devoted. Among them, HTS zeolite exhibits delightful catalytic performance, with $H_2O_2$ as an oxidant under mild conditions, as shown in **Figure 14**. It is apparent to illustrate that the cyclohexane conversion for HTS zeolite is as twice the time as that of TS-1 in the same reaction conditions, owing to the intensification of the accessibility of framework Ti species through hollow cavities. Except for the excellent conversion rate, HTS zeolite is also helpful for the high selectivity of cyclohexanone, by promoting the further oxidation of cyclohexanol with $H_2O_2$.

Particularly, it should be mentioned that the selectivity of KA oil is much higher than that of air-oxidation route, over 90% under optimized conditions. Although $H_2O_2$ solution is much more expansive than air, there is no need to carry out the complex separation and only few by-products formed, with higher atom utilization efficiency and thus meeting the principles of green chemistry. Consequently, this route provides an alternative method for KA oil production to replace the conventional route, and the scale up study is still under investigating

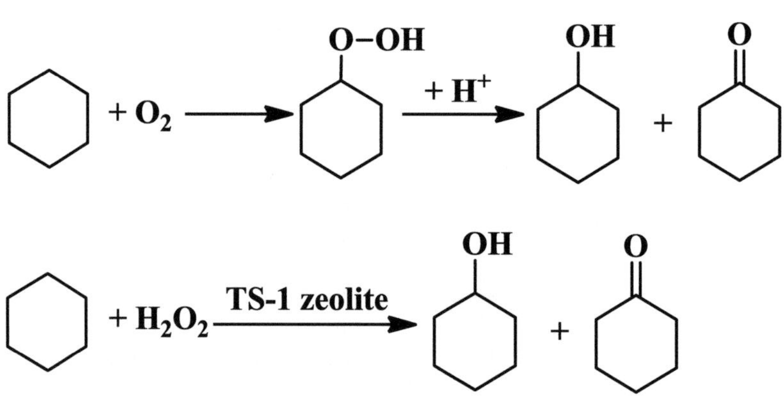


**Scheme 6.** Comparison of cyclohexane oxidation reaction network between commercial route and TS-1 catalyzed route with $H_2O_2$.

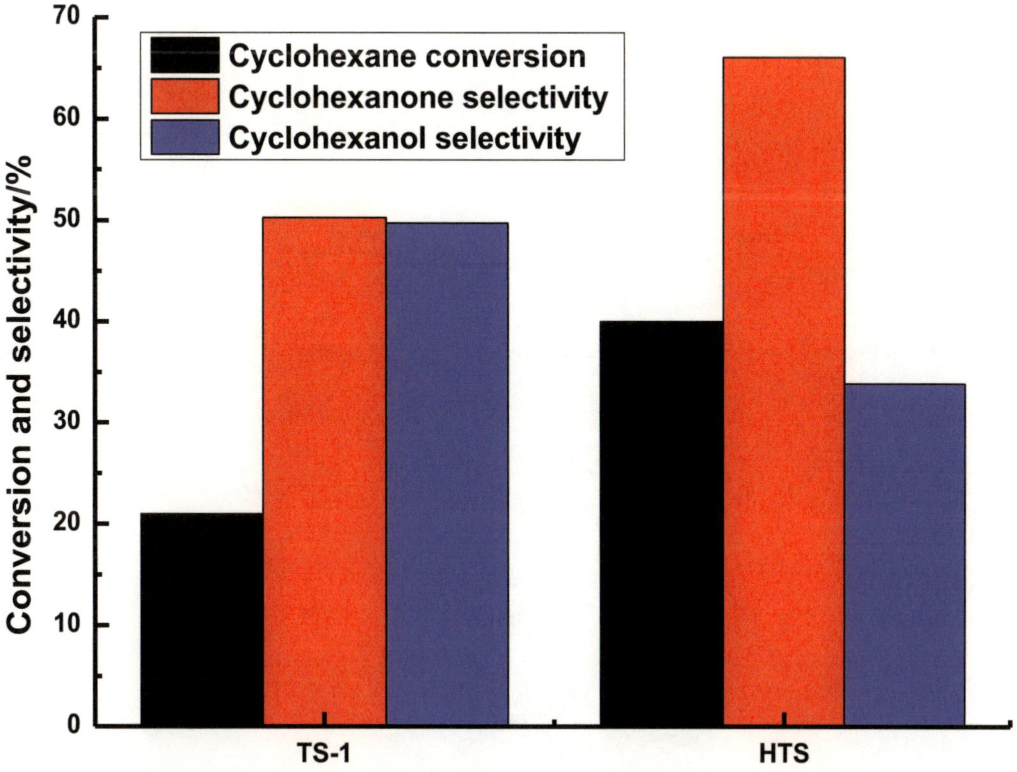


**Figure 14.** Catalytic performance of TS-1 and HTS in cyclohexane oxidation with $H_2O_2$ (reaction conditions: $H_2O_2$/cyclohexane = 1:1, 5 wt% catalyst content, 80°C, 500 rmp stirring).

at SINOPEC. We confirm that it is of great industrial potential, when taking the successful commercialization cases into consideration.

## 10. Epoxidation of fatty acid methyl ester with TBHP

As the fast developing of transformation of biomass to bulk and fine chemicals, green conversion processes for biomass utilization become particularly important [116–118]. In this context, epoxy fatty acid ester, one of the largest-scale chemicals derived from biodiesel, has been widely used as environment-friendly polyvinyl chloride (PVC) plasticizer and intermediates in the production of fine chemicals, such as nonanal, nonene, and other bulky molecules used as a biolubricant. In industry, epoxy fatty acid ester is adopted by the peracid process (Prileshajev reaction, **Schemes 7** and **8**). However, the Prileshajev method is of low efficiency, high risk, and pollution, due to the use of inorganic acid as catalyst and hazardous peracids as an oxidant.

Recently, we have developed a heterogeneous FAME epoxidation process with TBHP catalyzed by titanosilicate zeolite to replace the Prilezhaev method. While due to mass transfer

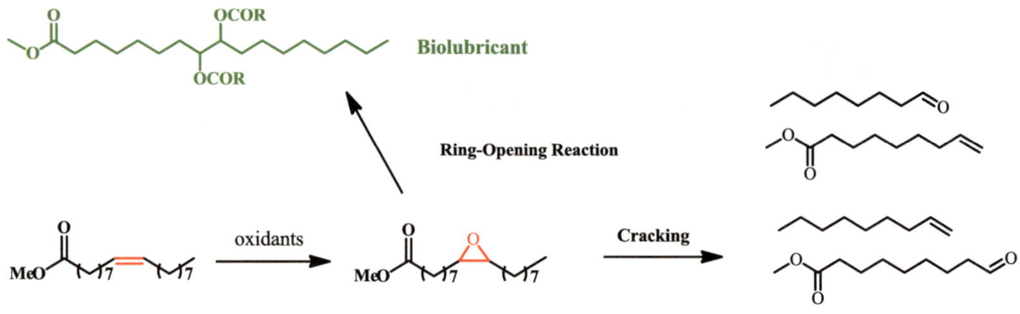


**Scheme 7.** FAME epoxidation process and side reactions catalyzed by TS-SP zeolite.

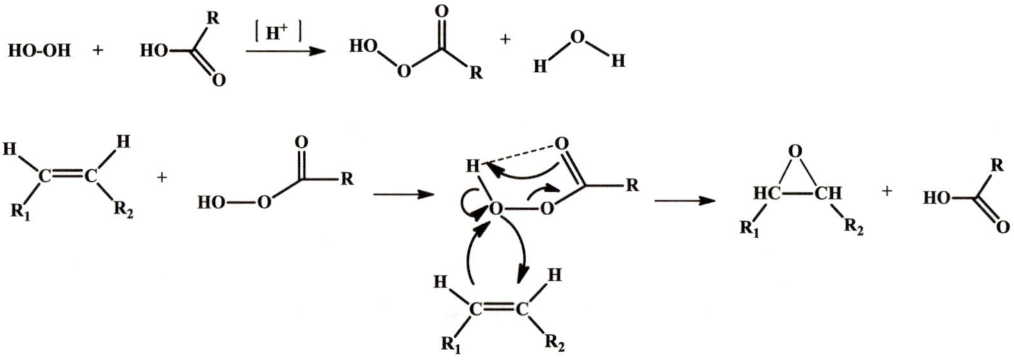

**Scheme 8.** Reaction pathways of the Prilezhaev reaction with $H_2O_2$ in industry.

limitation of FAME, TS-1 presents low activity with FAME conversion of 16.8% and a fairly low epoxy fatty acid ester at 0.5 h. In order to overcome mass transfer limitation in TS-1, several methods have been proposed. Recently, we have exploited a hierarchical titanosilicate zeolite with spongy structure (TS-SP) by combining pore-embedding and post-treatment methods. The specific surface area of TS-SP calculated by Brunauer-Emmett-Teller (BET) method (**Table 4**) is 518 $m^2 \cdot g^{-1}$, comparing with TS-1 of 435 $m^2 \cdot g^{-1}$, corresponding external surface area calculated by the t-plot method is 342 $m^2 \cdot g^{-1}$, which is also far more than traditional TS-1, indicating the existing of mesopores or macropores. Pore-size distribution and cumulative pore volume calculated by a nonlocal density functional theory (NL-DFT) model confirmed that obvious and abundant mesopores distributed in the range of 3–5 nm were formed in the zeolite. Thus, in epoxidation of FAME, the uniform mesopores promote the diffusion

| Zeolites | Ti/Si ratio (%) | $S_{BET}$ ($m^2g^{-1}$) | $S_{ext}$ ($m^2g^{-1}$) | $V_{micro}$ ($cm^3g^{-1}$) | $D_{meso}$ (nm) |
|---|---|---|---|---|---|
| TS-1 | 3.58 | 435 | 8 | 0.181 | 2.5, 4.3 |
| TS-SP | 3.51 | 518 | 342 | 0.077 | 3–5.5 |

**Table 4.** Textural parameters and elementary composition of titanosilicate zeolites.

of FAME and TBHP molecules inside the zeolites, leading to the better accessibility of active sites and facilitating product transport outside the pores, as shown in **Figure 15**. As a result, the spongy zeolite shows even more remarkable catalytic activity with 91.5% conversion at 0.5 h, which is 5.4 times as that of TS-1. What is more, FAME could be completely consumed within only 3 h, with the epoxy product selectivity more than 90%, which is far higher than that of TS-1. To the best of our best knowledge, this result is close to the highest performance published in the literature so far.

In conclusion, heterogeneous catalytic epoxidation of FAME is of significant importance for the environmental protection, saving energy, and improving selectivity of target product, as it avoids the use of hazardous peracids and inorganic acids. Those achievements confirm that mass transformation limitation of FAME could be solved by introducing hierarchical pores with zeolite particles, bringing significant enhancement of catalytic activity in bulk molecule oxidation reactions.

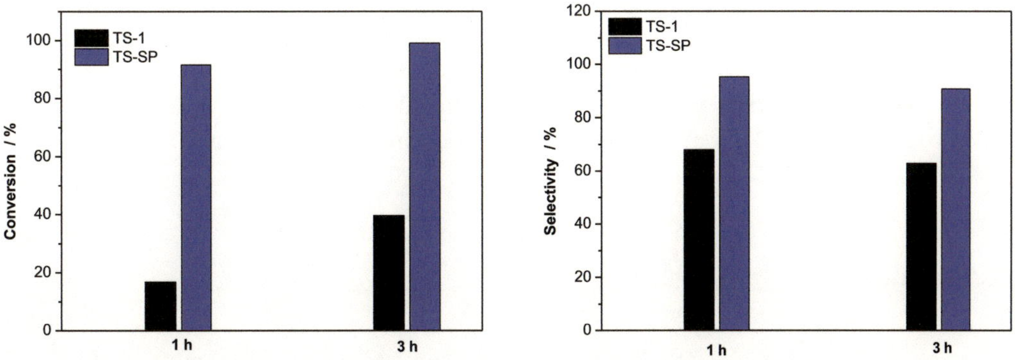


**Figure 15.** Catalytic performance of various TS-1 zeolites in epoxidation of methyl oleate with TBHP.

## 11. Conclusion and perspective

In summary, we have systematically reviewed the synthesis of hierarchical TS-1 zeolites and their catalytic application in several oxidation processes at SINOPEC, to replace the hazard and polluted traditional routes. In the early years, HTS zeolite was synthesized by using a "dissolution-recrystallization" treatment under hydrothermal conditions, which demonstrates much higher catalytic performance than conventional TS-1 zeolite, due to the diffusion improvement by intracrystalline confined mesopores. In addition, post-synthesis helps the reincorporation of Ti species into the framework matrix, thus HTS zeolite exhibits excellent activity in many commercial processes, such as propylene epoxidation, cyclohexanone ammoximation, and phenol hydroxylation. Moreover, we have already developed several interesting oxidations, such as chlorohydrination of AC with $H_2O_2$ and HCl and cyclohexanone BV oxidation with $H_2O_2$, for meeting the requirements of sustainable and green chemistry. On the other side, we also paid more attention on the exploitation of hierarchical TS-1 zeolites, for enhancing the entrance and mass diffusion with open mesopores, by using combined silanization and post-treatment technologies. Therefore, the application of TS-1 zeolites has been extended to the catalytic oxidation of bulk subtracts and oxidants (TBHP and CHHP), showing far better advantages in the epoxidation of FAME with TBHP over the Prileshajev reaction with $H_2O_2$. Most importantly, after long-term experiment and thinking, we can rationally design and synthesize hierarchical TS-1 zeolites with different kinds of mesopores at molecular scale, according to the demands for specific reactions. Furthermore, these catalytic materials provide bright opportunities for replacing traditionally polluted oxidation processes, and many successful cases have been achieved. For further development, many novel hierarchical zeolites with different framework composition and new topological structures will be prepared, and then these materials will also find many cleaner catalytic reactions, with an evident of reducing the generation of waste and unwanted products.

## Acknowledgements

This work was financially supported by the National Basic Research Program of China (973 Program, 2006CB202508) and the China Petrochemical Corporation (SINOPEC Group 20673054).

## Author details

Changjiu Xia[1], Xinxin Peng[1], Yao Zhang[1], Baorong Wang[2], Min Lin[1*], Bin Zhu[1], Yibin Luo[1] and Xingtian Shu[1]

*Address all correspondence to: minlin@tom.com

1 State Key Laboratory of Catalytic Materials and Reaction Engineering, Research Institute of Petroleum Processing, SINOPEC, Beijing, PR China

2 National Engineering Research Center for Carbohydrate Synthesis, Jiangxi Normal University, Nanchang, PR China

## References

[1] Nemeth L, Bare SR. Science and technology of framework metal-containing zeotype catalysts. Advances in Catalysis. 2014;**57**:1–97

[2] Reddy JS, Kumar R, Ratnasamy P. Titanium silicalite-2: Synthesis, characterization and catalytic properties. Applied Catalysis. 1990;**58**(1):1–4

[3] Fejes P, Nagy JB, Kovács K, et al. Synthesis of tin (IV) silicalites (MFI) and their characterization: A Mössbauer and MAS NMR spectroscopy study. Applied Catalysis A: General. 1996;**145**(1–2):155–84

[4] Perez-Pariente J, Martens JA, Jacobs PA. Crystallization mechanism of zeolite beta from $(TEA)_2O$, $Na_2O$ and $K_2O$ containing aluminosilicate gels. Applied Catalysis. 1987;**31**(1):35–64

[5] Taramasso M, Perego G, Notari B. Preparation of porous crystalline synthetic material comprised of silicon and titanium oxides: U.S. Patent 4,410,501. 1983-10-18

[6] Zhu Y, Chuah G K, Jaenicke S. Selective Meerwein-Ponndorf-Verley reduction of $\alpha,\beta$-unsaturated aldehydes over Zr-zeolite beta. Journal of Catalysis. 2006;**241**(1):25–33

[7] Boronat M, Corma A, Renz M. Mechanism of the Meerwein-Ponndorf-Verley-Oppenauer (MPVO) redox equilibrium on Sn– and Zr–beta zeolite catalysts. Journal of Physical Chemistry B. 2006;**110**(42):21168–21174

[8] Pérez-Ramírez J, Groen J C, Brückner A, et al. Evolution of isomorphously substituted iron zeolites during activation: Comparison of Fe-beta and Fe-ZSM-5. Journal of Catalysis. 2005;**232**(2):318–334

[9] Patra AK, Dutta A, Pramanik M, et al. Synthesis of hierarchical mesoporous Mn–MFI zeolite nanoparticles: A unique architecture of heterogeneous catalyst for the aerobic oxidation of thiols to disulfides. ChemCatChem. 2014;**6**(1):220–229

[10] Kim JH, Namba S, Yashima T. Para-selectivity of metallosilicates with MFI zeolite structure. Zeolites. 1991;**11**(1):59–63

[11] Boronat M, Concepción P, Corma A, et al. Determination of the catalytically active oxidation Lewis acid sites in Sn-beta zeolites, and their optimisation by the combination of theoretical and experimental studies. Journal of Catalysis. 2005;**234**(1):111–118

[12] Van der Waal JC, Lin P, Rigutto MS, et al. Synthesis of aluminium free titanium silicate with the BEA structure using a new and selective template and its use as a catalyst in epoxidations. Studies in Surface Science and Catalysis. 1997;**105**:1093–1100

[13] Wolf P, Hammond C, Conrad S, et al. Post-synthetic preparation of Sn-, Ti- and Zr-beta: A facile route to water tolerant, highly active Lewis acidic zeolites. Dalton Transactions. 2014;**43**(11):4514–4519

[14] Zhong W, Mao L, Yi W, et al. Highly efficient light-driven $HNO_3$ nitration-oxidation of cyclohexane to co-product nitrocyclohexane and adipic acid in a biphasic system. Research on Chemical Intermediates. 2016;**42**(2):461–470

[15] Hassall CH. The Baeyer-Villiger oxidation of aldehydes and ketones. Organic Reactions. 2011, **9**(3):73–106

[16] Van de Vyver S, Román-Leshkov Y. Emerging catalytic processes for the production of adipic acid. Catalysis Science & Technology. 2013;**3**(6):1465–1479

[17] Román-Leshkov Y, Moliner M, Labinger JA, et al. Mechanism of glucose isomerization using a solid Lewis acid catalyst in water. Angewandte Chemie International Edition. 2010;**49**(47):8954–8957

[18] Wu P, Tatsumi T, Komatsu T, et al. Postsynthesis, characterization, and catalytic properties in alkene epoxidation of hydrothermally stable mesoporous Ti-SBA-15. Chemistry of Materials. 2002;**14**(4):1657–1664

[19] Luo HY, Lewis JD, Román-Leshkov Y. Lewis acid zeolites for biomass conversion: Perspectives and challenges on reactivity, synthesis, and stability. Annual Review of Chemical and Biomolecular Engineering. 2016;**7**:663–692

[20] Cesana A, Mantegazza MA, Pastori M. A study of the organic by-products in the cyclohexanone ammoximation. Journal of Molecular Catalysis A: Chemical. 1997;**117**(1):367–373

[21] Fornasari G, Trifiro F. Oxidation with no-redox oxides: Ammoximation of cyclohexanone on amorphous silicas. Catalysis Today. 1998;**41**(4):443–455

[22] Reddy JS, Sivasanker S, Ratnasamy P. Ammoximation of cyclohexanone over a titanium silicate molecular sieve, TS-2. Journal of Molecular Catalysis. 1991;**69**(3):383–392

[23] Bellussi G, Millini R, Pollesel P, et al. Zeolite science and technology at Eni. New Journal of Chemistry. 2016;**40**(5):4061–4077

[24] Perego C, Carati A, Ingallina P, et al. Production of titanium containing molecular sieves and their application in catalysis. Applied Catalysis A: General. 2001;**221**(1): 63–72

[25] Zhou J, Hua Z, Cui X, et al. Hierarchical mesoporous TS-1 zeolite: A highly active and extraordinarily stable catalyst for the selective oxidation of 2,3,6-trimethylphenol. Chemical Communications. 2010;**46**(27):4994–4996

[26] Sanz R, Serrano DP, Pizarro P, et al. Hierarchical TS-1 zeolite synthesized from $SiO_2$-$TiO_2$ xerogels imprinted with silanized protozeolitic units. Chemical Engineering Journal. 2011;**171**(3):1428–1438

[27] Wang W, Li G, Li W, et al. Synthesis of hierarchical TS-1 by caramel templating. Chemical Communications. 2011;**47**(12):3529–3531

[28] Wang Y, Lin M, Tuel A. Hollow TS-1 crystals formed via a dissolution–recrystallization process. Microporous and Mesoporous Materials. 2007;**102**(1):80–85

[29] Lin M, Shu X, Wang X. Study on the synthesis and application of environmentally friendly catalyst TS-1 zeolite. Petroleum Processing and Petrochemicals. 1999;**30**:1–4

[30] Xie Z, Liu Z, Wang Y, et al. Applied catalysis for sustainable development of chemical industry in China. National Science Review. 2015;**2**(2):167–182

[31] Peng X, Xia C, Lin M, et al. Chlorohydrination of allyl chloride with HCl and $H_2O_2$ catalyzed by hollow titanium silicate zeolite to produce dichloropropanol. Green Chemistry. 2017;19:1221–1225

[32] Changjiu X, Bin Z, Min L, et al. A "Green" cyclohexanone oxidation route catalyzed by hollow titanium silicate zeolite for preparing ε-caprolactone, 6-hydroxyhexanoic acid and adipic acid. China Petroleum Processing & Petrochemical Technology. 2012;**14**(4):33–41

[33] Shi C, Zhu B, Lin M, et al. Propylene direct epoxidation performance of bi-functional titanosilicate with hollow structure. Catalysis Letters. 2009;**133**(1–2):70

[34] Lin M, Shu X, Wang X, et al. Titanium-silicalite molecular sieve and the method for its preparation: U.S. Patent 6,475,465. 2002-11–5

[35] Wang B, Lin M, Zhu B, et al. The synthesis, characterization and catalytic activity of the hierarchical TS-1 with the intracrystalline voids and grooves. Catalysis Communications. 2016;**75**:69–73

[36] Wang B, Lin M, Peng X, et al. Hierarchical TS-1 synthesized effectively by post-modification with TPAOH and ammonium hydroxide. RSC Advances. 2016;**6**(51):44963–44971

[37] Xia C, M Lin, B Zhu, Y Luo, X Shu. Hollow titanium silicalite zeolite: From fundamental research to commercial application in environmental-friendly catalytic oxidation processes. In: Zeolites-Useful Minerals. InTech, Croatia; 2016

[38] Serrano DP, Pizarro P. Synthesis strategies in the search for hierarchical zeolites. Chemical Society Reviews. 2013;**42**(9):4004–4035

[39] Serrano DP, Sanz R, Pizarro P, et al. Tailoring the properties of hierarchical TS-1 zeolite synthesized from silanized protozeolitic units. Applied Catalysis A: General. 2012;**435**:32–42

[40] Bordiga S, Bonino F, Damin A, et al. Reactivity of Ti (iv) species hosted in TS-1 towards H2O2–H2O solutions investigated by ab initio cluster and periodic approaches combined with experimental XANES and EXAFS data: A review and new highlights. Physical Chemistry Chemical Physics. 2007;**9**(35):4854–4878

[41] Guo Q, Sun K, Feng Z, et al. A thorough investigation of the active titanium species in TS-1 zeolite by in situ UV resonance Raman spectroscopy. Chemistry: A European Journal. 2012;**18**(43):13854–13860

[42] Zhang J, Tan J, Wang K, et al. Chlorohydrination of allyl chloride to dichloropropanol in a microchemical system. Industrial & Engineering Chemistry Research. 2012;**51**(45): 14685–14691

[43] Machado RM, Chapman TW. Kinetics and selectivity of aqueous propylene halohydrination. Industrial & Engineering Chemistry Research. 1989;**28**(12):1789–1794

[44] Luo ZH, You XZ, Li HR. Direct preparation kinetics of 1,3-dichloro-2-propanol from glycerol using acetic acid catalyst. Industrial & Engineering Chemistry Research. 2008;**48** (1):446–452

[45] Gao H, Lu G, Suo J, et al. Epoxidation of allyl chloride with hydrogen peroxide catalyzed by titanium silicalite 1. Applied Catalysis A: General. 1996;**138**(1):27–38

[46] Wang Q, Mi Z, Wang Y, et al. Epoxidation of allyl choride with molecular oxygen and 2-ethyl-anthrahydroquinone catalyzed by TS-1. Journal of Molecular Catalysis A: Chemical. 2005;**229**(1):71–75

[47] Russo V, Tesser R, Santacesaria E, et al. Chemical and technical aspects of propene oxide production via hydrogen peroxide (HPPO process). Industrial & Engineering Chemistry Research. 2013;**52**(3):1168–1178

[48] Xiong G, Cao Y, Guo Z, et al. The roles of different titanium species in TS-1 zeolite in propylene epoxidation studied by in situ UV Raman spectroscopy. Physical Chemistry Chemical Physics. 2016;**18**(1):190–196

[49] Lu Y, Wang R, Zhang J, et al. Evaluation of an improved epichlorohydrin synthesis from dichloropropanol using a microchemical system. Chinese Journal of Chemical Engineering. 2015;**23**(7):1123–1130

[50] Nakagawara S, Ozawa Y, Hotta K, et al. Spectroscopic characterization and the pH dependence of bactericidal activity of the aqueous chlorine solution. Analytical Sciences. 1998;**14**(4):691–698

[51] Yoon BH, Wang LJ, Yoon SL, et al. Mechanism of chlorate formation in chlorine dioxide delignification. Appita Journal: Journal of the Technical Association of the Australian and New Zealand Pulp and Paper Industry. 2004;**57**(6):472

[52] Geiger F M, Hicks JM, de Dios AC. Ab initio study of HOCl, HCl, $H_2O$, and $Cl_2$ interacting with four water molecules. The Journal of Physical Chemistry A. 1998;**102**(9):1514–1522

[53] Wu P, Komatsu T, Yashima T. Ammoximation of ketones over titanium mordenite. Journal of Catalysis. 1997;**168**(2):400–411

[54] Thangaraj A, Sivasanker S, Ratnasamy P. Catalytic properties of crystalline titanium silicalites III. Ammoximation of cyclohexanone. Journal of Catalysis. 1991;**131**(2):394–400

[55] Song F, Liu Y, Wang L, et al. Highly selective synthesis of methyl ethyl ketone oxime through ammoximation over Ti-MWW. Applied Catalysis A: General. 2007;**327**(1):22–31

[56] Chu CQ, Zhao HT, Qi YY, et al. Density functional theory studies on hydroxylamine mechanism of cyclohexanone ammoximation on titanium silicalite-1 catalyst. Journal of Molecular Modeling. 2013;**19**(6):2217–2224

[57] Zecchina A, Spoto G, Bordiga S, et al. Ammoximation of cyclohexanone on titanium silicalite: Investigation of the reaction mechanism. Studies in Surface Science and Catalysis. 1993;**75**:719–729

[58] Petrini G, Leofanti G, Mantegazza MA, et al. Caprolactam via ammoximation. 1996.

[59] Mantegazza MA, Petrini G, Spano G, et al. Selective oxidations with hydrogen peroxide and titanium silicalite catalyst. Journal of Molecular Catalysis A: Chemical. 1999;**146**(1):223–228

[60] Xuhong M, Dianzhong W, Yongrui W, et al. Nanosized molecular sieves as petroleum refining and petrochemical catalysts. Chinese Journal of Catalysis. 2013;**34**(1):69–79

[61] Wang Y, Zhang S, Zhao Y, et al. The mechanism of catalyst deactivation and by-product formation in acetone ammoximation catalyzed by hollow titanium silicalite. Journal of Molecular Catalysis A: Chemical. 2014;**385**:1–6

[62] Munakata H, Oumi Y, Miyamoto A. A DFT study on peroxo-complex in titanosilicate catalyst: Hydrogen peroxide activation on titanosilicalite-1 catalyst and reaction mechanisms for catalytic olefin epoxidation and for hydroxylamine formation from ammonia. The Journal of Physical Chemistry B. 2001;**105**(17):3493–3501

[63] Zecchina A, Bordiga S, Lamberti C, et al. Structural characterization of Ti centres in Ti-silicalite and reaction mechanisms in cyclohexanone ammoximation. Catalysis Today. 1996;**32**(1–4):97–106

[64] Xia C, Lin M, Zheng A, et al. Irreversible deactivation of hollow TS-1 zeolite caused by the formation of acidic amorphous $TiO_2$–$SiO_2$ nanoparticles in a commercial cyclohexanone ammoximation process. Journal of Catalysis. 2016;**338**:340–348

[65] Xia C, Lin M, Peng X, et al. Regeneration of deactivated hollow titanium silicalite zeolite from commercial ammoximation process by encapsulating amorphous $TiO_2$-$SiO_2$ nanoparticles inside zeolite crystal. Chemistry Select. 2016;**1**(14):4187–4192

[66] Zheng A, Xia C, Xiang Y, et al. Titanium species in deactivated HTS-1 zeolite from industrial cyclohexanone ammoxidation process. Catalysis Communications. 2014;**45**:34–38

[67] Meiers R, Hölderich WF. Epoxidation of propylene and direct synthesis of hydrogen peroxide by hydrogen and oxygen. Catalysis Letters. 1999;**59**(2):161–163

[68] Cumaranatunge L, Delgass WN. Enhancement of Au capture efficiency and activity of Au/TS-1 catalysts for propylene epoxidation. Journal of Catalysis. 2005;**232**(1):38–42

[69] Serrano DP, Sanz R, Pizarro P, et al. Preparation of extruded catalysts based on TS-1 zeolite for their application in propylene epoxidation. Catalysis Today. 2009;**143**(1):151–157

[70] Sharifov GS. Kinetics of chlorohydrination of allyl chloride. Kinetics and Catalysis. 2001;**42**(5):615–619

[71] Burgin J, Hearne G, Rust F. Derivatives of allylic chlorides. Industrial & Engineering Chemistry. 1941;**33**(3):385–388

[72] Kiryu T, Abe K, Honma A, et al. Chlorohydrination of allylchloride and methallylchloride. Journal of the Japan Petroleum Institute. 1969;**12**(5):350–355

[73] http://corporate.evonik.com/en/company/locations/location-asia/pages/skc-evonik-peroxide-korea-co-ltd.aspx

[74] http://www.chemicals-technology.com/projects/basf-hppo/

[75] Bassler P, Weidenbach M, Goebbel H. Chemical Engineering Transactions. 2010;**21:**571–576

[76] Lin M, Xia C, Zhu B, et al. Green and efficient epoxidation of propylene with hydrogen peroxide (HPPO process) catalyzed by hollow TS-1 zeolite: A 1.0 kt/a pilot-scale study. Chemical Engineering Journal. 2016;**295**:370–375

[77] Tanev PT, Chibwe M, Pinnavaia TJ. Titanium-containing mesoporous molecular sieves for catalytic oxidation of aromatic compounds. Nature. 1994;**368**(6469):321

[78] Mimoun H, Saussine L, Daire E, et al. Vanadium (V) peroxy complexes. New versatile biomimetic reagents for epoxidation of olefins and hydroxylation of alkanes and aromatic hydrocarbons. Journal of the American Chemical Society. 1983;**105**(10):3101–3110

[79] Thangaraj A, Kumar R, Ratnasamy P. Catalytic properties of crystalline titanium silicalites II. Hydroxylation of phenol with hydrogen peroxide over TS-1 zeolites. Journal of Catalysis. 1991;**131**(1):294–297

[80] Varagnat J. Hydroquinone and pyrocatechol production by direct oxidation of phenol. Industrial & Engineering Chemistry Product Research and Development. 1976;**15**(3): 212–215

[81] Neidig ML, Hirsekorn KF. Insight into contributions to phenol selectivity in the solution oxidation of benzene to phenol with $H_2O_2$. Catalysis Communications. 2011;**12**(6):480–484

[82] Suresh S, Srivastava VC, Mishra IM. Adsorption of catechol, resorcinol, hydroquinone, and their derivatives: A review. International Journal of Energy and Environmental Engineering. 2012;**3**(1):32

[83] Xia C, Long L, Zhu B, et al. Enhancing the selectivity of para-dihydroxybenzene in hollow titanium silicalite zeolite catalyzed phenol hydroxylation by introducing acid–base sites. Catalysis Communications. 2016;**80**:49–52

[84] Jiménez-Sanchidrián C, Ruiz JR. The Baeyer–Villiger reaction on heterogeneous catalysts. Tetrahedron. 2008;**64**(9):2011–2026

[85] Ten Brink GJ, Arends I, Sheldon RA. The Baeyer-Villiger reaction: New developments toward greener procedures. Chemical Reviews. 2004;**104**(9):4105–4124

[86] Krow GR. The Baeyer-Villiger oxidation of ketones and aldehydes. Organic Reactions. 2004;**43**(3):251–798

[87] Strukul G. Transition metal catalysis in the Baeyer-Villiger oxidation of ketones. Angewandte Chemie International Edition. 1998;**37**(9):1198–1209

[88] Kotlewska AJ, van Rantwijk F, Sheldon RA, et al. Epoxidation and Baeyer-Villiger oxidation using hydrogen peroxide and a lipase dissolved in ionic liquids. Green Chemistry. 2011;**13**(8):2154–2160

[89] Figg TM, Webb JR, Cundari TR, et al. Carbon-oxygen bond formation via organometallic Baeyer-Villiger transformations: A computational study on the impact of metal identity. Journal of the American Chemical Society. 2012;**134**(4):2332–2339

[90] Renz M, Meunier B. 100 years of Baeyer-Villiger oxidations. European Journal of Organic Chemistry. 1999;**1999**(4):737–750

[91] Pande CS, Jain N. Polymer-supported persulfonic acid as oxidising agent. Synthetic Communications. 1989;**19**(7–8):1271–1279

[92] Renz M, Blasco T, Corma A, et al. Selective and shape-selective Baeyer-Villiger oxidations of aromatic aldehydes and cyclic ketones with Sn-beta zeolites and $H_2O_2$. Chemistry: A European Journal. 2002;**8**(20):4708–4717